Lecture Notes in Computer Science 10218

Commenced Publication in 1973
Founding and Former Series Editors:
Gerhard Goos, Juris Hartmanis, and Jan van Leeuwen

More information about this series at http://www.springer.com/series/7409

Ivana Podnar Žarko · Arne Broering
Sergios Soursos · Martin Serrano (Eds.)

Interoperability and Open-Source Solutions for the Internet of Things

Second International Workshop, InterOSS-IoT 2016
Held in Conjunction with IoT 2016
Stuttgart, Germany, November 7, 2016
Invited Papers

 Springer

Editors
Ivana Podnar Žarko
Faculty of Electrical Engineering and
 Computing
University of Zagreb
Zagreb
Croatia

Arne Broering
Siemens AG Corporate Technology
Munich
Germany

Sergios Soursos
Intracom SA Telecom Solutions
Peania
Greece

Martin Serrano
Insight Centre for Data Analysis
National University of Ireland Galway
Galway
Ireland

ISSN 0302-9743   ISSN 1611-3349   (electronic)
Lecture Notes in Computer Science
ISBN 978-3-319-56876-8   ISBN 978-3-319-56877-5   (eBook)
DOI 10.1007/978-3-319-56877-5

Library of Congress Control Number: 2017937146

LNCS Sublibrary: SL3 – Information Systems and Applications, incl. Internet/Web, and HCI

Printed on acid-free paper

This Springer imprint is published by Springer Nature
The registered company is Springer International Publishing AG
The registered company address is: Gewerbestrasse 11, 6330 Cham, Switzerland

Preface

This volume contains the proceedings of the Second Workshop on Interoperability and Open-Source Solutions for the Internet of Things (InterOSS-IoT) held in Stuttgart, Germany, on November 7, 2016, and co-located with the 6th International Conference on the Internet of Things (IoT 2016). The workshop was co-organized by the H2020 projects symbIoTe and BIG IoT, which are part of the Internet of Things European Platform Initiative (IoT-EPI) working on relevant aspects for enabling and bridging the gaps on IoT interoperability.

The evolution in the IoT domain has created a complex ecosystem populated by numerous platforms that provide access to a broad range of software objects (virtual) and real-world devices (physical) "things." IoT-deployed platforms typically promote their specific interfaces and information models generating technology fragmentation gaps, and most likely adopt non-standard, and sometimes fully proprietary, protocols to control a variety of things. This reflects poorly on platform interoperability and creates a number of open issues and gaps. Platform fragmentation and lack of interoperability between current IoT solutions prevents the emergence of cooperative IoT ecosystems where applications can be built to use things operated by various IoT platforms, or simply to enable things operated by different platforms to interact and exchange information.

The InterOSS-IoT workshop featured an invited keynote talk and 13 oral presentations of original scientific papers. The event attracted around 40 participants who contributed constructively to all discussions. A big highlight of the event was the talk of the keynote speaker Ralph Müller, who did not fail to inspire the audience by giving great insights on the role of open source in the IoT domain. The volume includes selected and extended papers presented at the workshop covering a wide range of aspects related to IoT interoperability, such as semantics, security, business models, and applications. All papers underwent a rigorous two-step review process so that the final selection of 11 papers is included in this volume out of 17 papers that were initially submitted for review.

We would like to express our gratitude to the keynote speaker, Mr. Ralph Müller, for his inspiring talk. The chairs' special thanks go to the Technical Program Committee members for their valuable efforts in the review process. We cordially thank all the authors for their contributions, the workshop participants for their interest and active involvement in the workshop program, as well as the organizers of the 6th International Conference on the Internet of Things (IoT 2016) for providing an excellent workshop venue. Thank you all for making this workshop a valuable experience.

February 2017

Ivana Podnar Žarko
Martin Serrano
Arne Broering
Sergios Soursos

Interoperability and Open-Source Solutions for the Internet of Things (InterOSS-IoT)

Second International Workshop,
Co-located with the 6th International Conference on the Internet of
Things (IoT 2016)
Stuttgart, Germany, November 7, 2016
Invited Papers
LNCS 10218

The workshop is co-organized by the H2020 projects **symbIoTe** and **BIG IoT**, which are part of the Internet of Things European Platform Initiative (IoT-EPI) working on relevant aspects for enabling and bridging the gaps on IoT interoperability.

Program Committee Co-chairs

Ivana Podnar Žarko	University of Zagreb, Croatia
Martin Serrano	INSIGHT Centre for Data Analytics, National University of Ireland Galway, Ireland
Arne Broering	Siemens AG, Germany
Sergios Soursos	Intracom Telecom, Greece

Publicity Chair

Sofia Aivalioti	Sensing and Control, Spain

Technical Program Committee

Alexander Gluhak	Digital Catapult, UK
Arkady Zaslavsky	CSIRO, Australia
Charalampos Doukas	CREATE-NET, Italy
Cosmin-Septimiu Nechifor	Siemens, Romania
Danh Le Phouc	TU Berlin, Germany
Erno Kovacs	NEC, Germany

Florian Michahelles   WoT Research Group, Siemens, Berkeley USA
Gianluca Insolvibile   Nextworks, Italy
Jean-Paul Calbimonte   HES-SO Valais-Wallis, Switzerland
Jelena Mitic   Siemens, Germany
John Soldatos   AIT, Greece
Kary Främling   Aalto University, Finland
Mirko Presser   Alexandra Institute, Denmark
Oscar Corcho   Universidad Politecnica de Madrid, Spain
Ovidiu Vermesan   SINTEF, Norway
Patricia Martigne   Orange, France
Payam Barnaghi   University of Surrey, UK
Prem Jayaraman   RMIT University, Australia
Reinhard Herzog   Fraunhofer IOSB, Germany
Srdan Krco   DUNAVNet, Serbia
Stefan Schmid   Bosch Corporate Research, Germany
Steffen Lohmann   Fraunhofer IAIS, Germany
Tajana Šimunic Rosing   University of California, San Diego, USA
Thomas Usländer   Fraunhofer IOSB, Germany

Contents

Platform Performance and Applications

Semantic Interoperability

Semantic Interoperability as Key to IoT Platform Federation

Michael Jacoby[1(✉)], Aleksandar Antonić[2], Karl Kreiner[3],
Roman Lapacz[4], and Jasmin Pielorz[3]

[1] Fraunhofer IOSB, Karlsruhe, Germany
`michael.jacoby@iosb.fraunhofer.de`
[2] Faculty of Electrical Engineering and Computing,
University of Zagreb, Zagreb, Croatia
`aleksandar.antonic@fer.hr`
[3] Austrian Institute of Technology (AIT), Vienna, Austria
`{karl.kreiner,jasmin.pielorz}@ait.ac.at`
[4] Poznan Supercomputing and Networking Center, Poznan, Poland
`romradz@man.poznan.pl`

Abstract. Semantic interoperability is the key technology to enable evolution of the Internet of Things (IoT) from its current state of independent vertical IoT silos to interconnected IoT platform federations. This paper analyzes the possible solution space on how to achieve semantic interoperability and presents five possible approaches in detail together with a discussion on implementation issues. It presents the H2020 symbIoTe project as an example on how semantic interoperability can be achieved using semantic mapping and SPARQL query re-writing. We conclude that the found approaches together with the proposed technologies have the potential to act as corner stone technologies for achieving semantic interoperability.

Keywords: Semantic interoperability · Internet of Things · IoT platform federation · Semantic mapping · SymbIoTe · SPARQL query re-writing

1 Introduction

Semantic Interoperability is the key to "data exchange and service creation across large vertical applications" as seen as next step of evolution of the IoT [8]. In order to enable building new innovative, applications which make use of data from multiple existing vertical IoT silos these systems must not only be able to exchange information but also have a common understanding of the meaning of this data. This means, even if today's IoT systems are willing to expose their data and resources to others their semantically incompatible information models become an issue to dynamically and automatically inter-operate as they have different descriptions or even understandings of resources and operational procedures. To enable dynamic and automated interoperability, new features like

© Springer International Publishing AG 2017
I. Podnar Žarko et al. (Eds.): InterOSS-IoT 2016, LNCS 10218, pp. 3–19, 2017.
DOI: 10.1007/978-3-319-56877-5_1

semantic annotation, well-defined semantic mapping, unified resource discovery and federated authentication and authorization are required which cannot solely be provided by the existing platforms on their own but rather need to be offered by some kind of interoperability framework mediating between the platforms. Moreover, in the era of virtualization and Any-as-a-Service models, there is a need to federate independent infrastructures and introduce simplified methods to provide virtual resources of different types and owners in a dynamic and consistent manner. Because semantic interoperability is the basis for building services addressing sophisticated requirements across heterogeneous vertical IoT platforms, in this paper we present our thoughts and concepts how semantic interoperability between IoT platforms can be achieved.

The remainder of the paper is organized as follows. Section 2 provides a definition of semantic interoperability as it is used in this paper as well as some background on semantic technologies. In Sect. 3 possible approaches to achieve semantic interoperability are presented on a rather abstract level and Sect. 4 gives some detailed insights on what to keep in mind when trying to realize them. Section 5 presents how the symbIoTe project is approaching semantic interoperability picking up one of the possible approaches introduced in Sect. 3 and showing how it is realized in symbIoTe. The paper closes with conclusions and future work in Sect. 6.

2   Related Work

In this section, we will provide a definition of the term semantic interoperability as used in this paper. As semantic interoperability is a compound word, we will first analyze existing definitions for each of the terms and from this conclude a definition of the whole term. Semantics, as seen in linguistics and philosophy, refers to the study of meaning which means the relation of signifiers like words, symbols or signs and their denotation [1]. In computer science, the meaning of semantics is basically the same, but here the relations of signifiers and their denotation need to be understandable and processable by machines. The most common way to achieve this is by using an ontology which is "an explicit specification of a [shared] conceptualization" [11] and can be imaged like a formally-defined information model.

In 2001, Tim Berners-Lee introduced the idea of the Semantic Web [2], proposing the evolution of the internet from a web of documents to a web of machine-readable and -understandable data, which is becoming more and more reality. The corner stone technologies of the Semantic Web are the Resource Description Format[1] (RDF), a lightweight (meta data) data model for describing ontologies, and SPARQL Protocol and RDF Query Language[2] (SPARQL), a query language for data in RDF format, which both are standardized by the World Wide Web Consortium (W3C).

---

[1] https://www.w3.org/RDF/.
[2] https://www.w3.org/TR/rdf-sparql-query/.

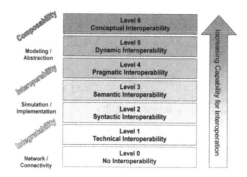

Fig. 1. The Levels of Conceptual Interoperability Model (from [25]).

"Broadly speaking, interoperability can be defined as a measure of the degree to which diverse systems, organizations, and/or individuals are able to work together to achieve a common goal" [14]. As this definition of interoperability is to broad in this context we also refer to the *Levels of Conceptual Interoperability Model* (LCIM) [25] depicted in Fig. 1. LCIM were created in the context of simulation theory but have a much broader applicability. In the scope of this paper, we see interoperability only up to Level 3 of LCIM where Level 1 refers to the low-level technical connectivity of platforms, Level 2 to using a common data format or protocol like XML or HTTP and Level 3 to having a unified understanding of the shared data. Therefore, semantic interoperability is defined as "the ability of computer systems to exchange data with unambiguous, shared meaning" [18] within this paper.

3 Approaches to Semantic Interoperability of IoT Platforms

The question discussed in this section is what possible approaches a system can take to achieve semantic interoperability between multiple IoT platforms. Figure 2 visually depicts the current situation where multiple platforms, in this case IoT platform A and B, having their own internal information model exist in parallel. To enable interoperability between those platforms they need to have a mutual understanding of things, i.e. some unification of their internal information models must somehow be defined.

As depicted in Fig. 3 the solution space to this problem can be thought of as a line between the two most radical approaches which are using a single core information model every platform must comply to on the one side and to not provide one at all and let platforms provide only their own information models which need to be aligned using semantic mapping on the other side. In between there exists a large, not clearly defined number of intermediate solutions from which three are representatively presented in the following together with the two radical approaches. These approaches are motivated by and in line with concepts presented by Wache and Choi et al. [4,28].

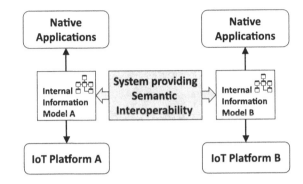

Fig. 2. Schematic representation of the problem of semantic interoperability between different IoT platforms.

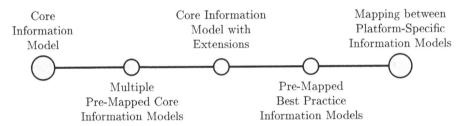

Fig. 3. Solution space for possible approaches to semantic interoperability.

3.1   Core Information Model

The most widespread approach amongst existing platforms is to use a single core information model that all platforms must comply with. This means that a platform can only expose data that fits into this core information model as custom extensions are not permitted. If a platform needs to expose data that does not fit into the core information model the platform cannot expose this data and cannot inter-operate with others.

Pros

- easy to implement and use since the data from all platforms follows the same information model
- resulting system easy to use for app developers who only need to know one information model

Cons

- defining an information model all platforms can agree upon may be difficult
- information model tends to become complex as it must comprise all data that should be exchangeable between platforms
- will always exclude some platforms whose internal information model does not fit the core information model

- no way to integrate future platforms with information models not compatible to the core information model without breaking the existing system

3.2   Multiple Pre-mapped Core Information Models

Based on the Single Core Information Model approach, this one tries to make it more easy and convenient for platform owners to integrate their internal information model by supporting not only a single core information model but multiple ones. To achieve that, a large number of existing platforms can easily participate it would be a good idea to choose well-established information models (e.g. the Semantic Sensor Network Ontology [5] or the oneM2M ontology [19]) as core information models. To ensure interoperability between platforms using different core information models the supported core information models are already mapped to each other. As it will not always be possible to map two core information models completely there will be some degree of information loss if platforms conform to different core information models but if they conform to the same one they will be fully interoperable.

Pros

- flexible approach as further core information models and mappings can be added over time
- does not enforce use of one single core information model which excludes less platforms from participating

Cons

- may still exclude some platforms whose information model does not match any of the core information models

3.3   Core Information Model with Extensions

This approach is based on an information model that is designed to be as abstract as possible but at the same time as detailed as needed. Therefore, the core information model should try to only define high-level classes and their interrelations which act as extension points for platform-specific instantiations of this information model. These platform-specific instantiations either use the provided classes directly or they can define a subclass which can hold any platform-specific extensions to the core information model, e.g. additional properties. Besides the high-level classes, the core information model may also contain properties the system needs which will be very general properties like *ID* or *name* in most of the cases. This results in an information model that has a minimalistic core that all platforms must conform to and extension points to realize custom requirements. Two platforms using different extensions can directly understand each other in terms of the core information model. When they need also to understand the custom extensions they must define a semantic mapping between their extensions.

Pros

- provides basic interoperability between platforms through minimalistic core information model
- provides full flexibility through custom extensions, i.e. no platforms are excluded
- high acceptance from adopter-side as it combines basic out-of-the-box interoperability (by the core information model) with support for complex scenarios (through extensions and semantic mapping)

Cons

- requires semantic mapping when custom extensions need to be understood by different platforms
- defining a semantic mapping can be a complex task and requires additional work from developers/platform owners
- design of the core information model is a complex task

3.4   Pre-mapped Best Practice Information Models

Essentially, this is the same approach as *Multiple Core Information Models* but with one small but significant modification: the provided information models are no longer seen as *core* information models but rather as *best practice* information models. Hence, platforms do not have to be compliant to any of the provided information models as in the previous approach but can choose their information model freely. If they choose to re-use one of the provided best practice information models they will gain instant interoperability to other platforms also aligned with one of the best practice information models.

Pros

- no limitations on information model, hence does not exclude any platform
- easier usage for platform owners
- better and broader interoperability due to already aligned best practice information models

Cons

- no initial interoperability between platforms as long as no mapping is defined when no pre-mapped information model is used
- defining a semantic mapping can be a complex task and requires additional work from developers/platform owners

3.5   Mapping Between Platform-Specific Information Models

In this approach, there isn't anything like a core information model. Instead, every platform independently provides its own information model. Interoperability is only achieved through mapping between these platform-specific information models.

Pros

- supports all possible information models and therefore all platforms
- mappings can be added iteratively increasing the degree of interoperability

Cons

- no initial interoperability between platforms as long as no mapping is defined
- defining a semantic mapping can be a complex task and requires additional work from developers/platform owners
- the system does not understand any of the data it is processing

4 Considerations About Realizing the Approaches

When thinking about realizing one of the above approaches there are multiple things that need to be considered. On the one hand, there are design decisions to make regarding the concrete specification of the information model(s). On the other hand, there are practical issues to take into account like what kind of software is needed to implement the chosen approach and what tools do already exist. This chapter will present two issues relevant for all approaches using semantic mapping followed by some approach-specific issues and a discussion on performance and scalability of the different approaches.

4.1 Semantic Mapping

Semantic mapping is used in four out of the five approaches as an alternative to defining a common information model all platforms can agree on. Figure 4 depicts a schematic representation how this could look like when implemented. At first, a platform owner must know that another platform he would like to inter-operate with exists. This issue is discussed in detail in the next section. For now, we assume that the platform owner of platform A knows that platform B exists and that he wants to define a mapping between the information models of the two platforms. To define such a mapping, which consists of multiple correspondence patterns and is called an alignment, he essentially needs a mapping language to express the alignment in. As defining such a mapping is not a trivial task, some tool support in form of a graphical alignment tool, i.e. a visual editor for the mapping language, is desirable. Optionally, to further ease the complexity of the task, a matcher could be up-streamed to automatically provide an initial mapping of the two information models. At run-time the mapping/alignment together with both information models is used by a mediator to translate instances between the two information models.

When trying to realize any of the approaches which include semantic mapping, these are areas that need to be analyzed for existing tools that fit the requirements. The mapping language is the most important component in this tool stack as all the other components need to be able to understand or generate mappings expressed in that language. The main criteria for choosing a language is its expressiveness and support for defining complex mappings.

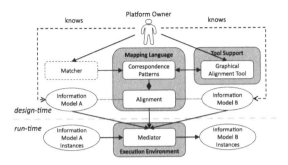

Fig. 4. Schematic representation of an example usage of semantic mapping.

4.2 Finding Other Platforms of Interest

A key problem when trying to make multiple platforms exposing their data in different information models interoperable using semantic mapping is the following: How to know that other platforms exist and that they provide data that is of such an interest that it justifies the effort to define a semantic mapping to it? As the parts of the information models that need to be mapped are platform-specific, there can be no semantic-based discovery but only a syntactical one as only the platform itself understands its information model. This implies that there is a need to have humans in the loop to close the semantic gap.

Therefore, a suitable candidate for enabling platform owners to find other platforms that might be of interest to them would be some kind of search functionality. Such a search could be quite primitive, e.g. a simple full-text search on the terms defined in the information model, or more sophisticated using concepts like phonetic search, natural language processing or translations to enable finding of relevant terms even if they do not syntactically match the search term or are expressed in another language.

Keeping in mind that mapping information models is key to enable interoperability between multiple heterogeneous platforms, this is an essential part of any system that aims to achieve this kind of interoperability and therefore needs to be solved appropriately.

4.3 Approach-Specific Issues

Single Core Information Model. The main challenge of this approach is to define a single core information model that does contain every information that any platform that should be made interoperable needs but, at the same time, to not make design decisions that prevent integration of upcoming platforms in the future. Since these two goals are contradicting this is an impossible task. Therefore, the single core information model approach is not suitable to provide interoperability between various kind of existing and upcoming IoT platforms. However, in a scenario where the domain is more narrow this approach could still be feasible.

Core Information Model With Extensions. Realizing this approach is essentially a trade-off between defining a quite detailed and easy-to-use core information model and finding the right level of abstraction to not make design decisions that exclude some platforms. This task is quite hard as, referring to the definitions provided by [24], the core information model is a hybrid between a domain ontology, defining only the very abstract structure of the IoT domain, and an application ontology, because it is especially tailored to be used with a single system. Therefore, this approach needs special attention on modeling the core information model to not bring platform-specific concepts, relations and properties into the core information model.

4.4 Performance and Scalability

In this Section, we analyze the above presented approaches regarding performance and scalability. Obviously, this strongly depends on the degree of dissimilarity of the information models of the platforms inter-operating. As long as all platforms are using the same information model all approaches will be able to perform well and to be scalable (depending on the implementation details of course). As for the Core Information Model approach this is given by definition. Therefore, it is the one with the best performance and scalability as all other approaches need to deal with semantic mapping. The Multiple Pre-Mapped Core Information Models approach has to cope only with a very limited number of possible different information models and therefore a limited number of mappings between them. This allows it to execute these mappings in a very optimized (probably hard-coded) way. Nevertheless, performance and scalability is worse than in the Core Information Model approach but can still be considered constant with number of interoperating platforms. The remaining three approaches are heavily based on user-defined mappings between information models. They have to discover and execute available mappings at run-time/query-time and therefore must provide a generic execution framework for generic mapping definitions. As the number of interoperating platforms, denoted n, grows over time, the number of mappings may rise up to n^2. Thus, for executing a query against all platforms up to $n - 1$ mappings need to be executed. When using query re-writing techniques for executing mappings, this may be done in parallel and therefore scalability is still given if the necessary computational resources are available. For these three approaches, performance will be slightly worse than with the Core Information Model approach and even than the Multiple Pre-Mapped Core Information Models as for each query the available mappings must the retrieved before execution. Also the execution of mappings will be slower than with the Multiple Pre-Mapped Core Information Models approach as the execution framework must be generic and cannot be optimized for a limited number of a priori known mappings.

5   SymbIoTe's Approach to Semantic Interoperability

SymbIoTe (**Symb**iosis of smart objects across **IoT e**nvironments) [23] is an EU project and part of the European Union's Horizon 2020 research and innovation programme. Its main objective is to provide an interoperability and mediation framework for collaboration and federation of vertical IoT platforms thus enabling creation of cross-domain applications using multiple heterogeneous IoT platforms in a unified way.

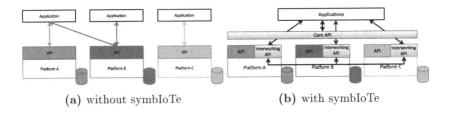

(a) without symbIoTe   (b) with symbIoTe

Fig. 5. IoT ecosystem with and without symbIoTe.

Figure 5a depicts the IoT ecosystem as it exists now. It consists of multiple IoT platforms that each represent a vertical silo which is tailored to a specific domain. If an application wants to integrate more than one platform, it has to do additional implementation work to make use of another platform-specific API. The vision of symbIoTe is to enable platform interoperability and creation of cross-platform apps between the existing vertical IoT silos with minimal integration effort for the platform owners as shown in Fig. 5b. Semantic interoperability is realized by the Core API providing a query functionality for meta data on available platforms and their resources. Syntactic interoperability is achieved by the Interworking API which provides a uniform access to resources of all platforms and can be seen as some kind of adapter that a platform owner needs to implement to be symbIoTe-compliant.

5.1   General Approach

Figure 7 shows how symbIoTe achieves semantic interoperability by implementing the Core Information Model with Extensions approach as presented in Sect. 3.3. On the left and the right, we see two existing IoT platforms exposing platform-specific APIs based on an internal information model to applications. Between those two vertical IoT stacks we see symbIoTe with the Core Information Model in its center. As proposed in Sect. 3.3, symbIoTe uses two central information models. The Core Information model describes domain specific information and the Meta Information Model describes symbIoTe internal meta data about platforms and resources. For a platform to become symbIoTe-compliant it must expose its data using a platform-specific information model which is basically the Core Information Model with platform-specific extensions

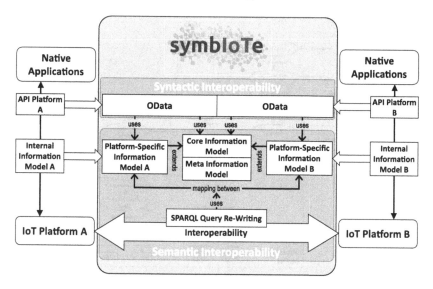

Fig. 6. High-level diagram showing how symbIoTe approaches syntactic and semantic interoperability.

to it. The main part of the actual interoperability happens via what is depicted at the arrow connection the two platform-specific information models: semantic mapping. This allows to define how the platform-specific extension of one platform can be translated into the platform-specific extensions of another platform thus allowing to define an arbitrary degree of interoperability between two platforms. When an app or a platform queries the Core API to find resources of interest on all available platforms, symbIoTe uses these mappings to re-write the query to fit the platform-specific information model of each platform and execute it against the meta data it has stored about each of them. Details on the symbIoTe Information Model as well as semantic mapping and SPARQL query re-writing are provided in the following sections.

5.2 SymbIoTe Information Model

The symbIoTe Information Model is comprised of two parts as depicted in Fig. 7. The first part is the Meta Information Model which covers all meta data about platforms that symbIoTe needs to store internally such as which platforms uses which Information Model and the URL of the Interworking API endpoint of a platform. Furthermore, it keeps track of the mappings between different information models that are described using any Mapping Language Information Model which will be explained in detail in the following section. The second part is the Core Information Model. Its design was driven by the trade-off to keep it is abstract as possible to not unnecessarily exclude any platform (as it may use another information model that doesn't fit the Core Information Model) but to include all information that symbIoTe needs to understand. This is due to the

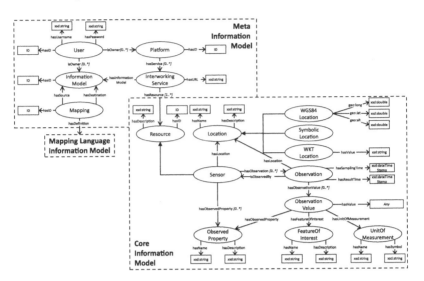

Fig. 7. SymbIoTe Information Model.

fact that symbIoTe internally can only understand the information coming from the platforms that is modeled within the Core Information Model but not the platform-specific extensions defined in the Platform-Specific Information Models. A good example for that is how locations are modeled. Initially, there was only *Location* defined and no sub-classes (which was thought to be defined in the platform-specific information models). But as symbIoTe needs to be able to internally understand the location to provide location-based query capabilities three sub-classes were defined to be used in the platform-specific information models so that symbIoTe can make use of the location information as it now is able to understand the meaning of this data.

Because this trade-off leaves only a very narrow solution space, symbIoTe cannot just re-use any existing ontology because their scope was either to abstract like the Semantic Sensor Network (SSN) ontology [5], thus resulting in symbIoTe not being able to understand the minimal information needed to work properly, or to narrow like the oneM2M Base Ontology [19], which would result in unnecessarily excluding platforms by over-specifying the information model.

For this reason, symbIoTe defines its own domain ontology for the IoT especially tailored to that narrow solution space to satisfy the trade-off between a desired high-level of abstraction and a needed concretization of some terms.

5.3   Semantic Mapping and SPARQL Query Re-writing

Semantic mapping and SPARQL query re-writing are closely related and together the most essential parts of providing semantic interoperability in symbIoTe. In this section we narrow down the general term *information model* used so far to refer to an information model realized as an ontology. Semantic mapping can therefore also be called ontology mapping in this section and is motivated by the fact that different IoT platforms may use different ontologies to describe their available resources covering (partially) the same domain and therefore could generally be (partially) interoperable. Even if these platform-specific ontologies essentially cover the same domain, they can describe this domain quite differently, e.g. use a taxonomy with another scope or granularity or use another terminology. These differences are called ontology mismatches and there exist multiple classifications for them [15, 20, 22, 26, 27]. Based on which of these types of mismatches a system should be able to resolve it needs to choose or develop a mapping language that offers language constructs to resolve these mismatches. There exist multiple existing mapping languages as OWL [17], C-OWL [3], MAFRA [16], SWRL [13], the Alignment Format [9] and EDOAL [7, 10] which differ widely about the supported kinds of mismatches. At the current state, symbIoTe is using EDOAL, the Expressive and Declarative Ontology Alignment Language, which is the most expressive of them and supports many different mismatch types [21, 22].

Having defined mappings/alignments between the different platform-specific information models/ontologies the platforms use to describe their available resources we don't instantly gain anything. Rather we need to implement some logic to access data presented in these different models in a unified way. Therefore, we need some execution environment with some kind of mediator like depicted in Fig. 4. For realizing such a mediator there are again two approaches which are essentially a trade-off between storage space needed and query execution time. The first one is to translate the actual data based on the mappings so that all data is stored according to each information model. This is very efficient with respect to query execution time but adds a massive overhead regarding storage space needed, especially when the number of different information models is high. The second approach is to store the data only according to its original information model and to re-write each query based on the existing mappings. This does not need any additional storage space but adds an overhead for multiple query translations. As the query can be translated and executed for each m apping in parallel symbIoTe uses this approach as shown in Fig. 6. In order to perform SPARQL query re-writing we are using the Mediation Toolkit[3] whose functionality is explained in detail here [6].

The overall algorithm for executing a SPARQL query formulated based on the platform-specific ontology of one IoT platform against the data of all platforms is comprised of three steps:

1. find all platforms and their interworking services for which a mapping exists from the ontology the query is formulated in (see Meta Information Model depicted in Fig. 7)

---

[3] https://github.com/correndo/mediation.

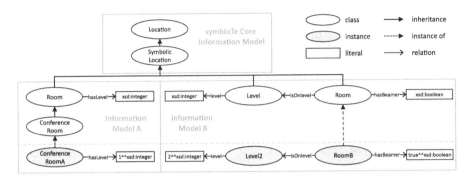

Fig. 8. Two example platform-specific information models which have some common concepts.

2. for each of the found mappings
   - re-write original SPARQL query based on the mapping
   - execute it against stored information about available resources
   - transform results back to match the ontology the query was originally formulated in
3. collect and return results

Example. In this section we present a simple example how semantic interoperability could be achieved using symbIoTe with semantic mapping and SPARQL query re-writing. Figure 8 shows two information models, A and B, which reflect parts of two PIMs in symbIoTe. As such, they both extend the symbIoTe Core Information Model, namely by adding subclasses to the *SymbolicLocation* class. Obviously, they both describe rooms and some of their properties in quite a similar but slightly different way. To enable the two platforms using these PIMs to exchange informations about rooms there needs to be a semantic mapping defined. Figure 9a shows the important excerpt of such a semantic mapping defined in the EDOAL language [7,10]. It contains two mapping rules. The first one (line 7–17) defines that a *ConferenceRoom* from model A equals a *Room* from model B with the property *hasBeamer* set to true. The second rule (line 19–24) defines that the relation *hasLevel* from model A is the same as following the relation *isOnLevel* followed by *level* of model B. Based on this mapping file, we can rewrite the SPARQL query shown in Fig. 9b written against model A to match model B using the Mediation Toolkit. The resulting query is depicted in Fig. 9c (note that the query has been manually updated by the authors to be more easy to read, the actual semantic of the query has not been changed).

As symbIoTe is currently work in progress and these are still areas of research, the chosen mapping language and SPARQL query re-writing framework might change in the future.

```
1    @prefix : <http://knowledgeweb.semanticweb.org/heterogeneity/alignment#> .
2    @prefix edoal: <http://ns.inria.org/edoal/1.0/#> .
3    @prefix modelA: <http://www.example.org/ontologyA#> .
4    @prefix modelB: <http://www.example.org/ontologyB#> .
5    @prefix xsd: <http://www.w3.org/2001/XMLSchema#> .
6
7    <http://www.example.com/mappings/plA-plB#Rule1> a :Cell ;
8        :entity1 modelA:ConferenceRoom ;
9        :entity2 [ a edoal:Class ;
10              edoal:and ( modelB:Room [ a edoal:AttributeValueRestriction ;
11                          edoal:onAttribute modelB:hasBeamer ;
12                          edoal:comparator edoal:equals ;
13                          edoal:value [ a edoal:Literal ;
14                              edoal:string "true" ;
15                              edoal:type "http://www.w3.org/2001/XMLSchema#boolean" ] ] ) ] ;
16        :measure "1.0"^^xsd:float ;
17        :relation "Equivalence" .
18
19   <http://www.example.com/mappings/plA-plB#Rule2> a :Cell ;
20        :entity1 modelA:hasLevel ;
21        :entity2 [ a edoal:Property ;
22              edoal:compose ( modelB:isOnLevel modelB:level ) ] ;
23        :measure "1.0"^^xsd:float ;
24        :relation "Equivalence" .
```

(a) Excerpt of the EDOAL mapping file in RDF/N3 syntax.

```
PREFIX modelA: <http://www.example.org/ontologyA#>

SELECT ?room ?level WHERE {
    ?room a modelA:ConferenceRoom ;
        modelA:hasLevel ?level .
}
```

```
PREFIX modelB: <http://www.example.org/ontologyB#>
PREFIX xsd: <http://www.w3.org/2001/XMLSchema#>

SELECT ?room ?level
WHERE
    { ?room  a modelB:Room ;
            modelB:hasBeamer  "true"^^xsd:boolean ;
            modelB:isOnLevel [ modelB:level ?level ] .
    }
```

(b) original SPARQL query   (c) re-written SPARQL query

Fig. 9. Example for SPARQL query re-writing based on a EDOAL mapping.

6 Conclusions and Future Work

In this paper we introduced five possible approaches to achieve semantic inter-operability along with detailed considerations on problems and risks to keep in mind when trying to implement them. We further presented the symbIoTe project as an example how to achieve not only internal but external interoper-ability as introduced in [12]. Moreover, the details of how symbIoTe realizes the Core Information Model with Extensions approach using semantic mapping and SPARQL query re-writing as core technologies was shown.

As the symbIoTe project is currently work in progress please note that the implementation details and used frameworks are subject to change. Furthermore, the following issues regarding the symbIoTe Information Model will be addressed in the future:

- add support for other resource types (actuators, services),
- revise modeling of Observations with focus on Location and FeatureOfInterest with regard to mobile sensors, and
- better user management.

Analyzing symbIoTe's approach we conclude that further research in the area of mapping languages and SPARQL query re-writing is essential for creating an interoperability framework for IoT platforms as these techniques are needed in three out of five possible approaches introduced in Sect. 3.

Acknowledgement. This work is supported by the H2020 symbIoTe project, which has received funding from the European Union's Horizon 2020 research and innovation programme under grant agreement No 688156. The authors would like to cordially thank the entire symbIoTe consortium for their valuable comments and discussions.

References

1. Merriam-webster dictionary. http://www.merriam-webster.com/dictionary/ semantics. Accessed 23 Oct 2016
2. Berners-Lee, T., Hendler, J., Lassila, O., et al.: The semantic web. Sci. Am. **284**(5), 28–37 (2001)
3. Bouquet, P., Giunchiglia, F., Harmelen, F., Serafini, L., Stuckenschmidt, H.: C-OWL: contextualizing ontologies. In: Fensel, D., Sycara, K., Mylopoulos, J. (eds.) ISWC 2003. LNCS, vol. 2870, pp. 164–179. Springer, Heidelberg (2003). doi:10. 1007/978-3-540-39718-2_11
4. Choi, N., Song, I.Y., Han, H.: A survey on ontology mapping. ACM Sigmod Rec. **35**(3), 34–41 (2006)
5. Compton, M., Barnaghi, P., Bermudez, L., et al.: The SSN ontology of the W3C semantic sensor network incubator group. Web Seman. Sci. Serv. Agents World Wide Web **17**, 25–32 (2012)
6. Correndo, G., Shadbolt, N.: Translating expressive ontology mappings into rewriting rules to implement query rewriting. In: International Semantic Web Conference, Ontology Matching Workshop. CEUR-WS (2011)
7. David, J., Euzenat, J., Scharffe, F.: The alignment API 4.0. Semant. Web **2**(1), 3–10 (2011)
8. European Commission: Reaping the full benefits of a digital single market. Comission Staff Working Document (2016)
9. Euzenat, J.: An API for ontology alignment. In: McIlraith, S.A., Plexousakis, D., Harmelen, F. (eds.) ISWC 2004. LNCS, vol. 3298, pp. 698–712. Springer, Heidelberg (2004). doi:10.1007/978-3-540-30475-3_48
10. Euzenat, J., Scharffe, F., Zimmermann, A.: Expressive alignment language and implementation. Knowledge Web project report, KWEB/2004/D2.2.10/1.0 (2007)
11. Gruber, T.R., et al.: A translation approach to portable ontology specifications. Knowl. Acquisition **5**(2), 199–220 (1993)
12. Herzog, R., Jacoby, M., Podnar Žarko, I.: Semantic interoperability in IoT-based automation infrastructures. AT - Automation Technology: Methods and Applications of Control, Regulation, and Information Technology (2015)
13. Horrocks, I., Patel-Schneider, P.F., et al.: SWRL: a semantic web rule language combining OWL and RuleML. W3C Member Submission **21**, 79 (2004)
14. Ide, N., Pustejovsky, J.: What does interoperability mean, anyway? toward an operational definition of interoperability for language technology. In: Proceedings of the Second International Conference on Global Interoperability for Language Resources, Hong Kong, China (2010)
15. Klein, M.: Combining and relating ontologies: an analysis of problems and solutions. In: IJCAI-2001 Workshop on Ontologies and Information Sharing, USA, pp. 53–62 (2001)
16. Maedche, A., Motik, B., Silva, N., Volz, R.: MAFRA — a mapping framework for distributed ontologies. In: Gómez-Pérez, A., Benjamins, V.R. (eds.) EKAW 2002. LNCS (LNAI), vol. 2473, pp. 235–250. Springer, Heidelberg (2002). doi:10.1007/ 3-540-45810-7_23

17. McGuinness, D.L., Van Harmelen, F., et al.: OWL web ontology language overview. W3C Recommendation **10**(10), 2004 (2004)
18. Network-Centric Operations Industry Consortium: Systems, capabilities, operations, programs, and enterprises (scope) model for interoperability assessment
19. oneM2M Partners: oneM2M base ontology. Technical report TS-0012-V2.0.0, oneM2M Partners, August 2016
20. Rebstock, M., Janina, F., Paulheim, H.: Ontologies-based business integration. Springer Science & Business Media (2008)
21. Scharffe, F.: Correspondence patterns representation. Ph.D. thesis, University of Innsbruck (2009)
22. Scharffe, F., Zamazal, O., Fensel, D.: Ontology alignment design patterns. Knowl. Inf. Syst. **40**(1), 1–28 (2014)
23. Soursos, S., Žarko, I.P., Zwickl, P., Gojmerac, I., Bianchi, G., Carrozzo, G.: Towards the cross-domain interoperability of IoT platforms. In: 2016 European Conference on Networks and Communications (2016)
24. Studer, R., Benjamins, V.R., Fensel, D.: Knowledge engineering: principles and methods. Data Knowl. Eng. **25**(1), 161–197 (1998)
25. Tolk, A., Wang, W., Wang, W.: The levels of conceptual interoperability model: applying systems engineering principles to M&S. In: Proceedings of the 2009 Spring Simulation Multiconference, p. 168 (2009)
26. Visser, P.R., Jones, D.M., Bench-Capon, T.J., Shave, M.J.: Assessing heterogeneity by classifying ontology mismatches. In: Proceedings of the FOIS, vol. 98 (1998)
27. Visser, P.R., Jones, D.M., Bench-Capon, T.J., Shave, M.: An analysis of ontology mismatches; heterogeneity versus interoperability. In: AAAI 1997 Spring Symposium on Ontological Engineering, Stanford CA, USA, pp. 164–172 (1997)
28. Wache, H., Voegele, T., Visser, U., Stuckenschmidt, H., Schuster, G., Neumann, H., Hübner, S.: Ontology-based integration of information-a survey of existing approaches. In: IJCAI-01 Workshop: Ontologies and Information Sharing, vol. 2001, pp. 108–117. Citeseer (2001)

Overcoming the Heterogeneity
in the Internet of Things for Smart Cities

Aqeel Kazmi$^{(\boxtimes)}$, Zeeshan Jan, Achille Zappa, and Martin Serrano

Insight Centre for Data Analytics, National University of Ireland, Galway, Ireland
{aqeel.kazmi,zeeshan.jan,achille.zappa,martin.serrano}@insight-centre.org
https://www.insight-centre.org

Abstract. In the past few years, the viability of the Internet of Things (IoT) technology has been demonstrated, leading to increased possibilities for novel human-centric services in the smart cities. This development has resulted in numerous approaches being proposed for harnessing IoT for smart city applications. Having received a significant attention by the research community and industry, IoT adaptation has gained momentum. IoT-enabled applications are being rapidly developed in a number of domains such as energy management, waste management, traffic control, mobility, healthcare, ambient assisted living, etc. On the other hand, this high-speed development and adaptation has resulted in the emergence of heterogeneous IoT architectures, standards, middlewares, and applications. This heterogeneity is hindrance in the realization of a much anticipated IoT global eco-system. Hence, the heterogeneity (from hardware level to application level) is a critical issue that needs high-priority and must be resolved as early as possible. In this article, we present and discuss the modelling of heterogeneous IoT data streams in order to overcome the challenge of heterogeneity. The data model is used within the VITAL project which is an open source IoT system of systems. The main objective of the VITAL platform is to enable rapid development of cross-platform and cross-context IoT based applications for smart cities.

Keywords: Internet of Things (IoT) · Interoperability · Linked data · Data model · Smart cities

1 Introduction

With an ever increasing urban population, governments are under pressure to manage resources efficiently while improving human-centric services. Governments and corporations alike are looking for ways to use Information and Communication Technologies (ICT) to provide sustainable solutions to the growing problems originating from rapid urbanization. The Internet of Things (IoT) is seen as a core technology that will help the governments and corporations to manage resources efficiently while improving human-centric services in the modern smart cities. Due to numerous benefits, the Internet of Things (IoT) has

© Springer International Publishing AG 2017
I. Podnar Žarko et al. (Eds.): InterOSS-IoT 2016, LNCS 10218, pp. 20–35, 2017.
DOI: 10.1007/978-3-319-56877-5_2

gained a significant amount of attention from the research community. In addition to research community efforts, serious business decisions taken by numerous major ICT companies such as, Google, Apple, Samsung, and Cisco have transferred the Internet of Things from conceptualization to reality [42]. Today the IoT, with an envisage of 25 billion connected devices by 2020, is seen as a prominent technology that can help in efficient resource management across different sectors such as smart energy management, waste management, smart traffic control, mobility management, smart healthcare, and Ambient-Assisted Living (AAL), etc. [41]. Without doubt, the Internet of Things has paved the way for smart cities to deliver cyber-physical based, context aware, human-centric services.

Attention given by the research community and industry has already led to the development of a number of heterogeneous IoT applications, which are offering services in different domains in the area of smart cities. In IoT, interoperability is seen as the process of integrating different levels of data (generated by an IoT application) that may also use different representation models. IoT systems that generate a set of heterogeneous data streams are unable to communicate with each other at the data level. The work presented in this paper aims to bring interoperability at one common layer by using semantics for storing heterogeneous data streams generated by different IoT eco-systems. This is achieved by means of a common data model using Linked Data technologies. In order to provide interoperability for multiple IoT data streams, we present system agnostic data models that are based on existing ontologies. The data models are used within the so-called Virtualized programmable InTerfAces for innovative cost-effective IoT depLoyments in smart cities (VITAL) project. VITAL uses linked data standards for modelling and accessing data including RDF as a basic data model, JSON-LD as the data format, and ontologies to specify the data in a formal way.

The remainder of the article is structured as follows. First we give a brief overview of the semantic and linked data technologies that are used to develop the data model. Then we present different ontologies, as well as necessary extensions to them for modelling data within VITAL, e.g. for modelling sensors and their measurements, for IoT systems and services, and for Smart City applications. Finally, we conclude the work and present future work plans.

2   IoT Fragmentation

In the recent past, IoT domain has made noteworthy progress across several dimensions, which includes, development of multiple IoT architectures and standards, IoT cloud-enabled middleware platforms, and a large number of IoT deployments in Smart Cities.

2.1   Architectures and Standards

Without doubt, the development of IoT architectures and related technological standards have provided the momentum for deploying scalable, intelligent, and interoperable IoT applications in the modern smart cities. A number of standardization organizations promoted these IoT architectures. Examples include the IoT-A Reference Model (ARM) [2]; the WWW Consortium's (W3C) Semantic Sensor Networks (SSN) incubator group that targeted the context-aware Wireless Sensors [32]; the Research Cluster on the Internet of Things (IERC) which aims to coordinate and build a consensus on ways to realise the IoT vision in Europe [8]; the European Commission's Alliance for Internet of Things Innovation (AIOTI) which aims to assist IoT research and standardization policies [1]; the Open Geospatial Consortium (OGC) which is committed to making open standards for the global geospatial community [18]; and the Electronic Product Code (EPCglobal) which focuses in two areas the RFID and Information Services [6]. The oneM2M standard [26], aims to develop technical specifications that can address the need for a common M2M Service Layer that can be readily embedded within various hardware and software. OM2M [19] is an open source implementation of the oneM2M standard.

2.2 Platforms

A number of middleware platforms have been developed, which facilitate the collection of data from homogeneous and heterogeneous IoT devices. Currently, available platforms provide a range of functionalities e.g. collecting data from hardware sensors, transforming it into a specified representation, and enabling (information access) interfaces for applications. Furthermore, some platforms also use the power of other ICT technologies, notably cloud computing infrastructures, and have opened ways towards IoT/cloud convergence. These platforms encourage the end-users to attach their IoT devices to the cloud infrastructure and enable easy-to-use APIs for retrieving sensor observations and developing applications [14,27]. HomeOS [36] is a platform that provides PC-like abstraction over a wide range of IoT devices. ThingWorx [28] is another IoT platform that facilitates the development of Smart City applications and has already been adopted by many corporations. As part of the IoT/cloud convergence, the OpenIoT (Open Source cloud solution for Internet of Things) is an open-source middleware platform that gathers sensor data and uses the cloud computing mechanism to offer Sensing as a Service [40] model. Examples of other cloud-enabled platforms include the Future Internet of Things (FIT) [9], Xively [33], and Hi-Reply [11] platforms. Similar to VITAL, BIG IoT is another EU project that aims at enabling the emergence of cross-standard, cross-platform, and cross-domain IoT services and applications [3]. However, VITAL differs from the BIG IoT, as BIG IoT does not transform or store the data streams coming from IoT systems.

2.3 Applications

The recent past has witnessed the development of many research and commercial IoT applications, which range from the very basic to more smart and intelligent ones [4,38,39]. These applications are typically developed for specific domains within the scope of modern Smart Cities and often use a large number of IoT devices. Examples of such applications include the BURBA (Bottom Up selection, collection and management of URBAn waste) system an Intelligent Waste management system [7]; the OpenIoT uses cases targeting smart manufacturing and smart agriculture [40]; OnFarm which is an IoT system designed to facilitate smart farming [15]; HeatWatch which is animal monitoring system that keeps track of animal movement [10]; SmartStructures a smart infrastructure monitoring solutions [25], ParkSight a smart parking management system within smart cities [4]; and Echelon which is a smart lighting management system targeting energy efficiency [24]. By looking at these examples, one can imagine that IoT applications are already transforming people's lives while having a significant impact on Smart Cities resource management.

2.4   Smart City Silos

The development of multiple IoT architectures, IoT platforms, and multiple (domain-specific) IoT applications has clearly instigated the momentum of IoT adaptation in Smart Cities. But on the other hand this development has introduced a significantly fragmented IoT landscape resulting from heterogeneity. Apart from development of architectures and platforms, this fragmentation has also resulted from independent IoT deployments, thereby leading to vertical IoT silos. These silos result from technological as well as organizational considerations. Horizontal convergence of these isolated IoT systems is essential to acquire new services and required efficiency [34]. This convergence should not be restricted to the technical integration rather it should also extend to the applications and services spread across different business contexts. IoT Data streams from the different domains should be combined to better manage the city services. For example, real-time information about the traffic flow combined with other information (weather, events, school timings, etc.) can enable traffic prediction based on time scales.

3   VITAL - System of Systems

VITAL has researched for best practices to eliminate the technological and organizational silos in the smart cities. It has introduced an integrated virtualized paradigm for the development, deployment and operation of smart city applications, which emphasizes the collection and processing of data streams from heterogeneous sensors and IoT platforms across the urban environment. This integrated virtualized paradigm is supported by VITAL. VITAL is a system of systems; a system that can support any underlying IoT system. VITAL has taken into account the work performed in the scope of the FP7 IoT-A project and their IoT-A Reference Model (ARM), as a guidance to design the VITAL architecture. The ARM has been used as a source of building blocks (e.g., protocols, interfaces, components), which can be used in order to assemble a concrete architecture. VITAL integrates several of these building blocks, with particular emphasis on

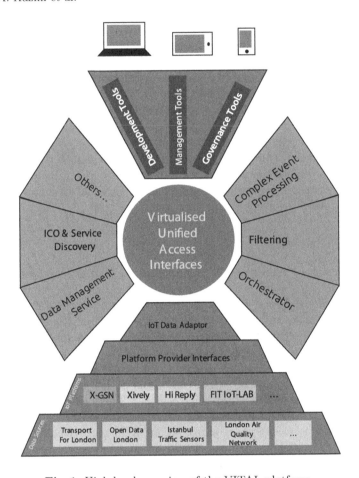

Fig. 1. High-level overview of the VITAL platform.

blocks that deal with services creation, orchestration and protocols, and less emphasis on low-level networking concepts and protocols. The development of VITAL is driven by a number of principles and characteristics:

- **Virtualization:** VITAL has developed mechanisms to support vitualized access to data generated by multiple IoT platforms and applications.
- **Modularity:** VITAL consists of a number of modules as depicted in Fig. 1. These components are developed as separate modules which can be deployed independently on different machines.
- **Standards-based:** VITAL uses a number of popular standards for data modelling e.g. RDF, JSON-LD, and ontologies. VITAL implements the Service Oriented Architecture (SOA) and enables RESTful web services for communication interchange mechanism.
- **Loosely Coupled:** VITAL in its nature is a service oriented distributed system which is developed around a loosely coupled approach.

– **Open Source**: VITAL is developed using open source technologies. To enable wide adaptation and maximize the impact, VITAL is distributed as an open source software under the LGPL license.

Figure 1 provides high-level architectural view of VITAL. Here we only provide an overview of VITAL and its components as the main focus of this article is the data models used in VITAL. At the lowest layer VITAL integrates multiple IoT systems and data sources (generated in multiple domains). The next layer is the Platform Provider Interfaces (PPI) that enables the access to metadata and data of IoT systems that are integrated. PPIs are responsible for mapping the data streams generated by IoT systems into VITAL data model for storage. The IoT Data Adaptor keeps track of registered PPIs, pulls data from IoT systems and stores it into the Data Management System (DMS). The DMS stores IoT data streams while enabling interfaces (for other components) to access stored (meta) data. VITAL platform also includes a number of added value functionalities such as Discovery Service for discovering IoT resources (e.g. sensors, systems, services, etc.), Filtering to filter information from DMS, Complex Event Processing (CEP) for event detection in IoT data streams, and Orchestrator for creating business specific services. Finally, the platform provides tools to allow the development of innovative cross-platform and cross-context applications and management and governance of IoT resources.

4   VITAL Data Model

Due to diverse set of use cases, VITAL covers a wide area of data models. It specifies the modelling of IoT systems and IoT services, e.g. using the Minimal Service Model (MSM) and basic IoT sensors and sensor measurements, e.g. using the Semantic Sensor Network (SSN) ontology. It also identifies data models for Smart City applications, e.g. smart transport systems. And it models the metadata for the VITAL system and its components, e.g. to model security and monitoring information. Next section provides an overview of the Linked Data technologies used in VITAL. Next we discuss existing ontologies and data models that are used as the basis of the VITAL data models and extended as needed by platform. This work is subdivided into different areas; sensors and sensor measurements, IoT systems and services, and Smart City Applications. We now list the ontologies we use for modelling the required data and also discuss the modelling of necessary additional data items.

4.1   Linked Data and Semantic

Linked Data (using RDF and ontologies) helps to describe and integrate data that is provided by different organisations in an interoperable way [37]. This is ideally suited for VITAL. VITAL envisages that a multitude of (independent) organisations and entities deploy and operate different sensors and IoT systems, which produce data (and offer services) that should be integrated in a system agnostic way. VITAL uses Linked Data standards for modelling and accessing metadata and data.

RDF: The Resource Description Framework (RDF) is World Wide Web Consortium (W3C)'s recommendations [22], which is also the most commonly used data model in the context of Linked Data. The RDF data model is a popular standard for describing things (known as resources or entities). By itself it is a graph-based data model that represents information as labelled directed graphs. This graph is built of triples that describe the data. Each triple (sub, pre, obj) consists of a subject sub, a predicate pre and an object obj. Using RDF in a Linked Data context has numerous advantages [37]:

- If the identifiers of data items (both used as subjects and objects) and vocabulary terms (used as predicates) are HTTP URIs, the RDF data model can be used at global scale and anybody is able to refer to anything.
- Each RDF triple is included in the Web of Data and can be the starting point for explorations in the data space, because any URI can be looked up in an RDF graph over the Web.
- It is possible to set RDF links between data from different sources.
- Sets of triples can be merged in a single graph to combine information.
- Terms taken from different vocabularies can be mixed in a RDF graph.

JSON-LD: A number of data formats are available that can be used to write RDF data, either directly as triples or as nodes that can be mapped to RDF triples, e.g. RDF/XML [22], RDFa [23], Turtle [30], N-Triples [21], and JSON-LD [12]. JSON-LD, a W3C recommendation, is a JSON-based serialization for Linked Data with the goals of simplicity, compatibility, terseness, expressiveness, etc. JSON-LD uses few important keywords, such as @context, @id, @value, and @type. In the VITAL platform JSON-LD is used. JSON-LD allows for referencing external files to provide context. This means contextual information can be requested on-demand which makes JSON-LD better suited to situations with high response times or low bandwidth usage requirements. Using JSON-LD will reduce the complexity of VITAL development by making it possible to reuse a large number of existing tools and reduce the inherent complexity of RDF documents.

Ontologies: Another important Linked Data concept used in VITAL is ontology. To integrate all data, generated by one or different sources, there have to be some rules. Some rules determine how the RDF graph is to be built and how triples may be connected or not. These rules are given by ontologies. An ontology specifies formally the conceptualisation of a domain of interest. As the conceptualisation is formal, a computer can automatically reason on it. There are many ontologies that have already been developed. Reuse of existing ontologies is crucial. If an existing ontology within the domain of use does not meet all the requirements and some new data models arise they should be attached to the existing ontology. We require ontologies in different areas. First, VITAL needs to model sensors and sensor measurements, which are the basis of any IoT system. Second, VITAL models IoT systems and services that are integrated into the VITAL. Third, VITAL provides means to model entities that are relevant to Smart City applications. And finally, it provides ontologies to model the VITAL system itself.

4.2   Sensors and Measurement

A number of ontologies have been developed to model sensors and sensor observations. Most of the existing ontologies are domain specific. Some abstract and generic ontologies are developed to provide a conceptual framework for IoT systems, for example, the Semantic Sensor Network (SSN) ontology developed by the W3C Semantic Sensor Network Incubator [35]. In practice such ontologies must be combined with additional ontologies to define concrete instances of abstract concepts. For instance, while a generic sensor ontology may specify how to model a sensor and its measurements, additional definitions must be used to model a concrete location of sensor and time when an observation was made. SSN is the generic sensor ontology that the VITAL consortium has selected to be used in VITAL.

SSN: The Semantic Sensor Network (SSN) ontology [35] defines a conceptual framework for describing sensors and sensor observations. The SSN ontology can formally describe sensors in terms of their:

- accuracy and capabilities,
- observations,
- measurement method,
- operating and survival ranges, and
- deployments.

A sensor could be any object that observes, it can be an electronic object, a virtual object or even a human. The ranges are used in the definition of sensors conjoined with the performance of these sensors. The description of deployment includes the deployment lifetime as well as the sensing purpose of the deployed macro instrument.

Time: Temporal aspects are essential for any system addressing real world phenomena, e.g. smart city IoT systems. Timestamps can be used to describe when a sensor observation was taken or when it was transferred. Multiple readings can be ordered by the time of their occurrence. Users may specify or query for certain types of observations based on specific timeframe. To model this, VITAL provides an ontology for time as well as temporal properties and relations. A well-established ontology for this is OWL Time [29]. OWL Time allows describing of temporal properties and relationships. It also supports time intervals as well as durations, which are useful for example, when describing imprecise measurement times as well as complex event specifications.

Location: Location in the physical world is another basic concept that is modelled in VITAL. There is a multitude of different location models and ontologies available today, including geographical and symbolic location models. VITAL follows a practical approach to allow easy usage of the system while at the same time being flexible enough for advanced use cases. WGS84 [31] coordinates are used as the basic location model, since they are the de-facto standard for outdoor localisation using the GPS system. In addition, symbolic names are often used as locations. VITAL uses the Linked GeoData system to model more complex location concepts, including symbolic names, cell-based locations and inter-location relationships.

Measurement: Different properties in the VITAL data models represent physical magnitudes like length or weight. Each one of these properties should be associated with an unambiguous unit of measurement, e.g. metre or kg. There is currently no single accepted ontology to model units of measurements in linked data. A number of potential ontologies were found and four were chosen for detailed evaluation. VITAL chooses QUDT [20] ontology for units of measurements due to the impressive scope and amount of information available on each type as well as the reputation of the maintainer and sponsoring party. QUDT is also actively maintained, with the latest version that was released in September 2016.

Modelling Sensor

To model sensors, sensor measurements and their descriptions in VITAL, we reuse and extend the SSN ontology and the Delivery Context (DC) [5] ontologies. A sensor is modelled as a `VitalSensor`, a subclass of `ssn:Sensor` and `dcn:Device`. By using the SSN ontology, VITAL can immediately describe sensors in detail, including aspects like the properties that they observe, sensor locations, and sensor observations. The SSN ontology also allows to model non-functional aspects of a sensor, e.g. its accuracy or reliability, by adding a `ssn:hasMeasurementProperty` property to the sensor description that points to a `ssn:MeasurementCapability`. The DC ontology defines a `dcn:Device` as a class that represents a device in the delivery context. By using the DC ontology, VITAL can reuse a highly detailed set of ontologies describing many aspects of devices, including their software, their hardware as well as their networking.

In addition to the SSN and DC ontologies, VITAL defines an additional property for sensors, `hasLastKnownLocation`. This property is a sub property of `dul:lastLocation` as specified in the SSN ontology description. It links to a location, which is the last known location of the sensor. The property does not imply that this is the actual current location of the sensor. If the sensor is mobile, it could have moved to a new location after the description was created. If the property is not available in a sensor description, then the location of the sensor may not be known.

Note that the location of a sensor can be modelled with different types as specified before, e.g. as a `geo:Point` in case GPS coordinates are used. To be as flexible as possible, we use the generic `dul:Entity` class to represent all different location types here. It is taken from the DOLCE+DnS Ultralite (DUL) ontology and defines it as anything: real, possible, or imaginary, which some modeller wants to talk about for some purpose [17]. In addition to the basic description of a sensor, some VITAL services may require additional information. Properties and classes to model this information are described in next sections.

Modelling Sensor Measurements

Similarly to sensors, VITAL uses the SSN ontology to model sensor measurements. A measurement is modelled as an `ssn:Observation`. The observation contains a link to an observed property (using `ssn:observedProperty`) to specify what the observation is measuring. In addition, it specifies when the measurement was taken (`ssn:observationResultTime`), at which location (`dul:hasLocation`) in WGS84 format, the quality of the measurement (`ssn:observationQuality`), as well as the measured value (`ssn:observationResult`) with the unit of measurement specified with the QUDT ontology.

4.3   IoT Systems and Services

VITAL integrates existing IoT systems (e.g. deployed platforms) and allows clients (applications as well as VITAL system services) to access (meta) data and services of such systems. Currently there are four platforms for which example deployments are integrated as a proof of concept: X-GSN, Hi-Reply, INRIA FIT, and Xively. To integrate systems and work with them, VITAL needs a set of models to describe IoT systems and their services.

Modelling Systems

VITAL models an `IotSystem` as a subclass of `ssn:System` with a number of additional properties. An IoT system description always includes a basic set of properties that describe general aspects of the system, e.g. its operator. In addition, a system description may specify a set of IoT services that it offers. To describe general metadata about the system, VITAL supports three new properties: `status`, `operator`, and `serviceArea`. The status of an IoT system might change during its lifecycle, thus, VITAL compliant systems that want to expose their current operational stat must manage a virtual sensor of type `MonitoringSensor` (a sub class of `VitalSensor`). Furthermore, `OperationalState` specifies the operational state of a system. A number of states are defined in VITAL as sub classes of `OperationalState`: `Operational`, `StartingUp`, `Running`, `ShuttingDown`, and `Unavailable`. In addition to the metadata discussed so far, an IoT system may offer a set of IoT services to access its functionalities. To allow an IoT system to link to descriptions of provided IoT services, VITAL introduces a new property `providesService`.

Modelling Services

In VITAL, an IoT system does not only provide access to IoT data (e.g. sensor measurements) but may offer a set of distinct and heterogeneous IoT services. An IoT service may be generic, e.g. a service to discovery ICOs or to access filtered data, or application specific, e.g. a service to reserve a parking space in a Smart City IoT system. In fact, VITAL models all functionality that can be exposed by an IoT system and can be accessed and used by a client as an IoT service, including data access, e.g. reading a sensor measurement. VITAL therefore specifies a flexible data model to specify all different kinds of IoT services. There is currently no single, standardised way to model IoT services. Based on the related work discussed before, we decided to base VITAL's semantic IoT service model on existing work in the domain of web services.

As discussed before, VITAL aims at providing a semantic model that is generic – yet simple and minimal, reuses existing ontologies as much as possible and allows to link with an active community as well as other current projects. After careful consideration, we selected to use widely used Minimal Service Model (MSM) [13] as the basis of its IoT service modelling ontology. In the VITAL system an IoT service is modelled as a RESTful (web) service that is described by Linked Data using the MSM ontologies. This allows publishing a description of the IoT service that can e.g. be used for discovery or for automatic composition tasks.

IoT systems (integrated within VITAL platform) may provide configuration functionalities. In order to model these functionalities, `Configuration-Service` class is defined as a sub class of `msm:Service` along with two operations: `GetConfiguration` in order to access existing configurations and `SetConfiguration` to set new configurations.

An IoT system can allow VITAL to monitor a number of monitoring functionalities. For example, the status of an IoT system, the status of sensors that an IoT system manages, performance metrics of an IoT system, SLA parameters related to an IoT system, etc. These functionalities are exposed by a `MonitoringService` class a sub class of `msm:Service` with a number of operations: `GetSystemStatus`, `GetSensorStatus`, `GetSupportedPerformanceMetrics`, `GetPerformanceMetrics`, `GetSupportedSLAParameters`, and `GetSLAParameters`.

The VITAL platform can use both a pull and push based mechanism to obtain observations made by a sensor. An IoT system with various sensors can provide/support both mechanism by providing an observation service. An IoT system must support at least one of these two mechanisms in order to allow access to sensor observations. This service is modelled as `ObservationService` sub class of `msm:Service` with the following operations: `GetObservations`, `SubscribeToObservationStream`, and `UnsubscribeFromObservationStream`.

4.4   Smart Cities

The VITAL platform is under proof-of-concept validation using the two use cases focus on Smart Transport and Traffic Management and Smart Working. Therefore, in the following we discuss how to model data items and properties that are relevant for these two scenarios. Clearly, VITAL is not restricted to these two use case scenarios. A user who would like to use VITAL for other smart city aspects can do so by specifying additional ontology elements. Due to the nature of Linked Data, these additional elements can be added at any time without the need to redesign the system.

Modelling Cities

VITAL obtains the majority of its semantic information on cities from DBpedia, using the classic DBpedia dataset for most information with the option of using DBpedia live for information that updates more frequently. It is also encouraged to link real places and services in cities back to DBpedia to improve the amount of knowledge available. For example, while Camden Road would be modelled as an `otn:Road` as part of a smart transport system, it should also link to `dbpedia:Camden_Road`. As discussed, `ssn:observes` is used to specify a property of the real world a sensor is observing. The specification for modelling real instances of such properties are application domain specific and should be defined as required. For these two use cases we re-use the ontology of FP7 project OpenIoT [40].

Modelling Smart Transport

VITAL models transport infrastructure using a combination of ontologies. The core of these is the Ontology of Transportation Networks (OTN) [16]. This ontology allows easy modelling of a transport network graph with connections between infrastructure such as bus and train stations as well as events such as accidents and blocked passages. In order to model the traffic management system, VITAL describes a class `TrafficManagement` a sub class of `IoTSystem`. For the purpose of transport and traffic scenarios and use cases (specifically in Camden Town, London), VITAL models the following sub classes of `ssn:Property: BusArrival`, `RailArrival`, `TubeArrival`, and `AvailableBikes`. To describe general description of `Line`, VITAL supports two new properties. These properties are: `name` and `direction`

Modelling Smart Working

The structural development in advanced economies is influencing the change in working patterns with (increasingly) employees more likely to work on the move. Mobile workers require optimal working environments to be available at short notice and without any difficulties to use. Owners of these suitable environments require optimal occupancy. To meet these requirements, the smarter working application (powered by VITAL) introduced a number of sub classes of `ssn:Property`: `AvailableDesks` and `Availability`. Each available workspace in the system that meets the specified criteria is shown on a map and/or list. Additional information is also shown associated with each workspace e.g. anticipated air quality, temperature, humidity, and footfall in the requested time window and location etc. In order to model additional information, following classes are defined as sub classes of `ssn:Property`: `CarbonMonoxide`, `Ozone`, and `Footfall`. Should the additional information need to be added and modelled by introducing new classes, they can be added as required.

4.5   JSON-LD Definitions

Since Linked Data in VITAL is always formatted as JSON-LD we introduce some additional definitions (in a JSON-LD context section) that do not change the used ontology or the resulting RDF triples but align the JSON-LD representation more closely to normal JSON and thus makes it easier for developers to work with the data. All JSON keys that do not specify a prefix will be expanded to URIs in the VITAL ontology namespace. This results in more compact files with less clutter. Then we define that the key id will be mapped to a JSON-LD node identifier (`@id`). The node identifier is used to create the URI that is used as the subject in RDF triples. Similarly, specified `key` will be mapped to the JSON-LD keyword `@type`. This results in an RDF triple being created that specifies the RDF type of a node. To further simplify the JSON-LD file, the key name will be mapped to `rdfs:label` and the key description will be mapped to `rdfs:comment`. We then specify a number of prefixes that can be used in the JSON-LD description to reduce the length of keys by specifying them as so-called terms. All these mappings are completely transparent to developers and can be ignored by clients. They are only relevant if the JSON-LD file is mapped to RDF triples internally. Together, they reduce the complexity of the resulting JSON-LD file and make it both smaller and easier to read and understand for JSON developers. The resulted JSON-LD contexts are available at:

– http://vital-iot.eu/contexts/system.jsonld
– http://vital-iot.eu/contexts/service.jsonld
– http://vital-iot.eu/contexts/sensor.jsonld
– http://vital-iot.eu/contexts/measurement.jsonld

4.6   VITAL Ontology

In previous sections we discussed the VITAL data models e.g., the major classes, properties, and operations that are used to model IoT systems, services, sensors, observations, and smart city applications. Clearly, discussion on each concept is impossible. Figure 2 visualizes the main entities in VITAL ontology. VITAL ontology can be accessed at: http://vital-iot.eu/ontology/ns/ontology.owl.

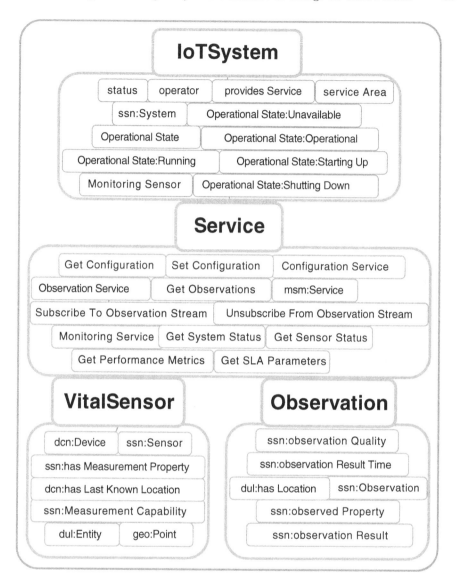

Fig. 2. VITAL data model: entities and their modelling

5   Conclusion and Future Work

The Internet of Things (IoT) domain has gained a significant attention from both academia and industry. This has led to the development of multiple heterogeneous architectures and standards, platforms, and IoT applications. Clearly, there is a pressing need that IoT applications address the challenge of heterogeneity and allow the exchange of information across platforms and applications.

This paper provides the basis for the semantic (meta) data models used in the VITAL system. We build upon Linked Data principles and technologies to provide interoperable and platform agnostic data models that are based on existing ontologies. This allows VITAL applications to integrate other data sources in the Web, resulting in a large and varied set of usable data items. Although we analysed a large number of ontologies during the design of the VITAL data models, the work is not finished. Thus, it is envisaged that more data models will be added as VITAL extends its functionalities and more VITAL-powered use cases are developed.

Acknowledgment. The authors acknowledge support from the ICT-2013.1.14 - VITAL-IoT project, which is co-funded by the European Commission under the Seventh Framework Program (FP7), grant agreement number 608682. The authors also acknowledge help and contributions from all partners of the VITAL consortium.

References

1. The alliance for internet of things innovation (aioti). http://www.aioti.eu. Accessed 4 Sep 2016
2. The architectural reference model (arm) - iot-a. http://www.iot-a.eu. Accessed 14 Sep 2016
3. Bridging the interoperability gap of the internet of things (big iot). http://big-iot.eu/. Accessed 10 Oct 2016
4. Connecting the real world: Parklight. Technical Report. Accessed 4 Sep 2016
5. Delivery context ontology. https://www.w3.org/TR/2009/WD-dcontology-20090616. Accessed 14 Sep 2016
6. The electronic product code (epcglobal). http://www.gs1.org/epcglobal. Accessed 4 Sep 2016
7. The eu-funded project burba. http://www.burbaproject.net. Accessed 4 Sep 2016
8. The european research cluster on the internet of things (ierc). http://www.internet-of-things-research.eu. Accessed 4 Sep 2016
9. Fit. https://www.fit-equipex.fr. Accessed 4 Sep 2016
10. Gea farm technologies: Heatwatch. http://www.jacesupplies.co.uk. Accessed 4 Sep 2016
11. Hireply. http://www.reply.eu. Accessed 4 Sep 2016
12. Json-ld 1.0: A json-based serialization for linked data. https://www.w3.org/TR/json-ld. Accessed 14 Sep 2016
13. Minimal service model. http://iserve.kmi.open.ac.uk/ns/msm/msm-2013-05-03.html. Accessed 14 Sep 2016
14. Netatmo. https://www.netatmo.com. Accessed 4 Sep 2016
15. The onfarm system: grow informed. http://www.onfarm.com. Accessed 4 Sep 2016
16. Ontology of transportation networks: Deliverable a1–d4, project rewerse: Reasoning on the web with rules and semantics. http://rewerse.net/deliverables/m18/a1-d4.pdf. Accessed 14 Sep 2016
17. Ontology:dolce+dns ultralite ontology. http://ontologydesignpatterns.org/wiki/Ontology:DOLCE+DnS_Ultralite. Accessed 14 Sep 2016
18. The open geospatial consortium (ogc). http://www.opengeospatial.org. Accessed 4 Sep 2016

19. Open source implementation of onem2m. http://www.eclipse.org/om2m. Accessed 11 Nov 2016
20. Qudt - quantities, units, dimensions and data types ontologies. http://qudt.org. Accessed 14 Sep 2016
21. Rdf 1.1 n-triples: A line-based syntax for an rdf graph. https://www.w3.org/TR/n-triples. Accessed 14 Sep 2016
22. Rdf 1.1 xml syntax. https://www.w3.org/TR/rdf-syntax-grammar. Accessed 14 Sep 2016
23. Rdfa 1.1 primer - third edition: Rich structured data markup for web documents. https://www.w3.org/TR/rdfa-primer. Accessed 14 Sep 2016
24. Smart street lighting. http://www.echelon.com. Accessed 4 Sep 2016
25. Smart structures: Smartpile. http://smart-structures.com. Accessed 4 Sep 2016
26. Standards for m2m and the internet of things. http://www.onem2m.org. Accessed 11 Nov 2016
27. Thingsspeak. https://www.thingspeak.com. Accessed 4 Sep 2016
28. Thingworx an iot platform. http://www.thingworx.com. Accessed 4 Sep 2016
29. Time ontology in owl. https://www.w3.org/TR/owl-time. Accessed 14 Sep 2016
30. Turtle - terse rdf triple language. https://www.w3.org/TeamSubmission/turtle. Accessed 14 Sep 2016
31. W3c geo positioning. https://www.w3.org/2003/01/geo/wgs84_pos. Accessed 14 Sep 2016
32. The w3c semantic sensor network (ssn) incubator group. https://www.w3.org/2005/Incubator/ssn. Accessed 4 Sep 2016
33. Xively. http://xively.com. Accessed 4 Sep 2016
34. International Electrotechnical Commission: Orchestrating infrastructure for sustainable Smart Cities (2014)
35. Compton, M., Barnaghi, P., Bermudez, L., García-Castro, R., Corcho, O., Cox, S., Graybeal, J., Hauswirth, M., Henson, C., Herzog, A., Huang, V., Janowicz, K., Kelsey, W. D., Phuoc, D. L., Lefort, L., Leggieri, M., Neuhaus, H., Nikolov, A., Page, K., Passant, A., Sheth, A., Taylor, K.: The {SSN} ontology of the {W3C} semantic sensor network incubator group. Web Semant. Sci. Serv. Agents World Wide Web **17**, 25–32 (2012)
36. Dixon, C., Mahajan, R., Agarwal, S., Brush, A., Lee, B., Saroiu, S., Bahl, P.: An operating system for the home. In: The 9th USENIX Symposium on Networked Systems Design and Implementation (2012)
37. Heath, T., Bizer, C.: Linked data: evolving the web into a global data space (2011)
38. Kazmi, A.H., O'Grady, M.J., Delaney, D.T., Ruzzelli, A.G., O'Hare, G.M.P.: A review of wireless-sensor-network-enabled building energy management systems. ACM Trans. Sen. Netw. **10**(4), 66: 1–66: 43 (2014)
39. Kazmi, A.H., O'Grady, M.J., O'Hare, G.M.P.: Energy management in the smart home. In: Proceedings of the 2013 IEEE 10th International Conference on Ubiquitous Intelligence & Computing and 2013 IEEE 10th International Conference on Autonomic & Trusted Computing (UIC-ATC 2013), pp. 480–486. IEEE Computer Society, Washington, DC (2013)
40. Serrano, M., Kefalakis, N., Soldatos, J., Hauswirth, M.: OpenIoT: an open source platform for enabling intelligence to the internet of things (2014)
41. Smith, I.G., Vermesan, O., Friess, P., Furness, A.: The Internet of Things 2012 New Horizons (2012)
42. Vermesan, O., Friess, P.: Internet of things - from research and innovation to market deployment (2014)

Interoperable Architectures and Platforms

An Architecture for Interoperable IoT Ecosystems

Stefan Schmid[1(✉)], Arne Bröring[2], Denis Kramer[3], Sebastian Käbisch[2],
Achille Zappa[4], Martin Lorenz[5], Yong Wang[6], Andreas Rausch[6], and Luca Gioppo[7]

[1] Robert Bosch GmbH, Stuttgart, Germany
stefan.schmid@bosch.com
[2] Siemens AG, Munich, Germany
{arne.broering,sebastian.kaebisch}@siemens.com
[3] Bosch Software Innovations GmbH, Stuttgart, Germany
denis.kramer@bosch-si.com
[4] Insight Centre for Data Analytics, NUI Galway, Galway, Ireland
achille.zappa@insight-centre.org
[5] Atos IT Solutions and Services GmbH, Vienna, Austria
martin.lorenz@atos.net
[6] Technical University Clausthal, Clausthal-Zellerfeld, Germany
{yong.wang,andreas.rausch}@tu-clausthal.de
[7] CSI-Piemonte, Torino, Italy
luca.gioppo@csi.it

Abstract. The Internet of Things (IoT) is maturing and more and more IoT platforms that give access to *things* are emerging. However, the real potential of the IoT lies in growing IoT cross-domain ecosystems on top of these platforms that will deliver new, unanticipated value added applications and services. We identified two crucial aspects that are important to grow an IoT ecosystem: (i) interoperability to enable cross-platform and even cross-domain application developments on top of IoT platforms as well as (ii) marketplaces to share and monetize IoT resources. Having these two crucial pillars of an IoT ecosystem in mind, we present in this article the *BIG IoT architecture* as the foundation to establish IoT ecosystems. The architecture fulfills essential requirements that have been assessed among industry and research organizations as part of the BIG IoT project. We demonstrate a first proof-of-concept implementation in the context of an exemplary smart cities scenario.

Keywords: Internet of Things · Architecture · Interoperability · Marketplace

1 Introduction

The idea of the Internet of Things (IoT) [1] has become more and more a commercial reality that spans various application domains, from smart homes, over smarter cities, to Industry 4.0. Various IoT platforms are upcoming: Cloud-level platforms such as

© Springer International Publishing AG 2017
I. Podnar Žarko et al. (Eds.): InterOSS-IoT 2016, LNCS 10218, pp. 39–55, 2017.
DOI: 10.1007/978-3-319-56877-5_3

Evrythng[1] or ThingWorx[2], and also on premise solutions such as Bosch's IoT Suite[3]. However, up to now, these IoT platforms failed to form vibrant IoT ecosystems. This is due to the large number of stakeholders, including developers and providers of platforms, services and applications. They require marketplaces that enable the monetization of their IoT resources. Once such marketplaces are established, revenue streams can be shared across all contributing stakeholders. A crucial task of a marketplace is to provide functionalities for advertising, discovery and orchestration of IoT services to facilitate their usage.

However, before such marketplaces can bring their effect, a serious market barrier needs to be tackled: the missing interoperability. The fragmentation of the IoT and the lack of interoperability prevent the emergence of broadly accepted IoT ecosystems [6]. A recent McKinsey study [2] estimates that a 40% share of the potential economic value of the IoT directly depends on interoperability between IoT platforms. Today, we are dealing with various vertically oriented and mostly closed systems. Architectures for IoT are built on heterogeneous standards (e.g., OMA LWM2M [3], OGC SWE [4] or OneM2M [5]) or even proprietary interfaces. This causes interoperability problems when overarching, cross-platform and cross-domain applications are to be built. Additionally, it leads to barriers for small innovative business, which cannot afford to offer their solution across multiple platforms.

In order to address these shortcomings in today's IoT landscape, this article concretizes our vision presented in [6]: It presents the BIG IoT architecture as enabler for establishing IoT ecosystems. It overcomes the above-described hurdles through (1) a common Web API, (2) semantic descriptions of resources and services, as well as (3) a marketplace as a nucleus of the ecosystem. We implement this architecture as part of the BIG IoT project[4]. This will allow new applications, e.g., by combining data from multiple platforms. In addition, platforms from multiple domains (e.g. home and city) and regions will be combined, such that applications can utilize all relevant information and work seamlessly across regions.

To ignite an IoT ecosystem based on the developed concepts, the BIG IoT project involves overall 8 IoT platforms. There are 6 cloud- or infrastructure-level platforms: Bosch's Smart City platform, based on the Bosch IoT Suite[3], CSI's Smart Data[5] platform, OpenIoT [7], Vodafone's Mobile Analytics Platform, VMZ's TIC[6] platform, and WorldSensing[7]. Further, there are 2 device-level platforms: Bosch's BEZIRK[8] platform and Econais' Wubby[9] platform. Together with multiple service and application implementations, these platforms will be BIG IoT-enabled and evaluated in 3 different pilots (Barcelona, Berlin/Wolfsburg and Piedmont).

---

[1] http://www.evrythng.com.
[2] http://www.thingworx.com.
[3] https://www.bosch-si.com/products/bosch-iot-suite/platform-as-service/paas.html.
[4] http://www.big-iot.eu.
[5] http://www.smartdatanet.it.
[6] https://viz.berlin.de/en/home.
[7] http://www.worldsensing.com.
[8] http://www.bezirk.com.
[9] http://www.wubby.io/.

The remainder of this article is structured as follows. Section 2 presents related work and outlines an overview of different research projects in this field. Section 3 outlines the high-level concepts and requirements for IoT ecosystems. Section 4 describes the BIG IoT realization of such an IoT ecosystem architecture, which is then demonstrated in a proof-of-concept in Sect. 5. Finally, we conclude this article in Sect. 6.

2 Background and Related Work

Various related works exist that contributed to the advancement of IoT architectures design. Most related to our work are other large research projects in context of the IoT. This section lists some of such approaches to give an overview of the research field and highlights the unique approach of our work in BIG IoT.

A prominent project in this context is the Internet of Things Architecture (IoT-A) project [8], which developed a comprehensive architectural reference model as a foundation for interoperability of IoT systems, including guidelines for the design of protocols and interfaces. However, other than IoT-A, which can be used as a blueprint for the development of an IoT platform, this work develops an architecture that focuses on integrating existing systems, components, and stakeholders of the IoT.

Another lighthouse IoT framework project is FI-WARE [9]. It develops a framework of so-called generic enablers to support IoT developments. Our approach differs from the FI-WARE idea, as we do not aim at creating another unified platform or platform building blocks, but enabling the coexistence and distributed collaboration of existing and already commercially deployed platforms to foster an easy creation of portable services by third party providers.

A Semantic Web-based design of a middleware platform for the IoT has been developed in the OpenIoT project [7]. While OpenIoT assumes the use of a single sensor middleware platform and its integration within a common cloud computing infrastructure, it does not address cross-platform mechanisms. This is however, a focal topic of the work described in this article. In fact, the OpenIoT platform is integrated into the BIG IoT project as one IoT platform of the overall ecosystem.

VITAL [10] aims at virtualized filtering and complex event processing mechanisms over a variety of IoT architectures. It focuses on an abstract virtualized digital layer that will operate across multiple IoT architectures. In that sense, VITAL has similar goals of integrating different IoT platforms. However, it is a domain specific effort, by restricting itself to smart cities. The project develops a centralized operating system, called Vital-OS, which manages and monitors all systems and data. In contrast, our work follows a domain-agnostic approach that generalizes emerging platforms and enables semantic interoperability to provide unified APIs.

The objective of CityPulse [11] has been to develop a distributed framework for the semantic discovery and processing of large-scale real-time IoT data and relevant social data streams for knowledge extraction in a city environment. CityPulse focuses on developments for the application layer. Their services could be integrated and run on top of the common API designed by BIG IoT.

The IoT@Work project [12] was a deep dive into the industry automation domain with its very specific requirements. The approach presented a novel solution for flexible production. The BIG IoT project in contrast aims at cross-domain applications and making use of existing platforms and installations in a more generic sense. By fostering the emergence of open ecosystems, our approach diverts from the specific one of IoT@Work.

One of the intentions of the BIG IoT project is to bring its approaches to standardization in order to reach an interoperable IoT platform landscape. In that sense, BIG IoT members are involved in the W3C Web of Things (WoT) [13] activities, which is going to be standardized in parallel to the BIG IoT project. The W3C WoT group was founded in spring of 2015, and the major motivation for initiating this group was also the fact that the IoT suffers from a lack of interoperability across platforms. BIG IoT members are mainly involved in the topic Thing Description. Development and experiences in WoT and BIG IoT are regularly synchronized in order to learn and benefit from each other.

Mineraud et al. [14] analyze the technological gaps of today's IoT platforms. Specifically, they highlight the fact that data and device catalogs as well as billing of consumers of the IoT data sources is generally missing. Hypercat [16] is an initiative that aims to address the issue of semantic interoperability for IoT through catalogs, which enable distributed data repositories to be used jointly by applications. As such, Hypercat offers a starting point to solving the issues of managing heterogeneous data sources through linked data and web approaches. However, the gap for IoT marketplaces, where providers and consumers of IoT resources can meet and do business, as they exchange their assets, are not yet addressed. Our work in the BIG IoT project targets to address exactly this vital gap. By building up on the previous works outlined above, we focus on reusing existing technologies with the goal of igniting IoT ecosystems.

3   High-Level Architecture Concepts and Requirements

This section first defines the terminology and key concepts for an IoT ecosystem architecture, then it identifies requirements from our stakeholders, deducts architectural implications, and high-level design decisions that influence and guide the design of the concrete BIG IoT architecture in Sect. 4.

3.1   Terminology and Conceptual Model for an IoT Ecosystem

Figure 1 defines the generic concepts that we identified within an IoT ecosystem and the interactions between them. The core concepts are: *offerings*, (offering) *providers* and *consumers*, and the interactions of *registering* and *discovering* offerings via a marketplace, and *accessing* the resources offered by a provider.

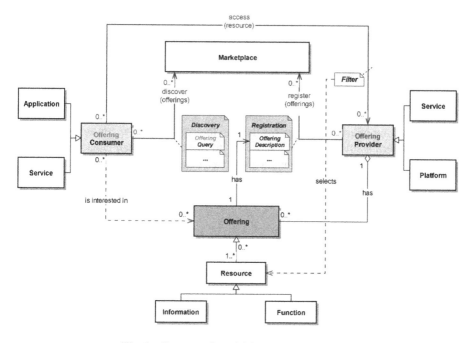

Fig. 1. Conceptual model for an IoT ecosystem.

An offering encompasses a set of IoT resources, typically a set of related information (e.g. low-level sensor data like temperature or aggregate information across an region) or functions (e.g. actuation tasks like open a gate or computational functions like compute a route), that are offered on a marketplace.

Providers register their offerings on a marketplace and provides access to the offered resources via a common API. A provider can be either a platform or a service instance that offers available resources, i.e., some information or access to functions that it wants to share or trade on the marketplace (e.g. an IoT platform of a parking lot provider). *Consumers* discover and subscribe to offerings of interest via a marketplace in order to access the resources. A consumer can be either an application or service instance that requires access to IoT resources in order to implement an intended service or function (e.g., a smart parking service provided by the city).

In technical terms, a provider registers its offerings on the marketplace by providing an offering description for each offering. An offering can for example entail parking information for a city and include data such as geo location or address of the parking lot, the type of lot (e.g. garage or on-street), available spots, occupied spots, etc. In order to increase interoperability between different IoT platforms, the offering description is provided in a machine interpretable manner, e.g., based on RDF [15] models. All relevant communication metadata is provided on how the offering can be accessed (e.g., endpoint URL, which HTTP method, etc.). As a default vocabulary set, the offering description includes a local identifier (unique to a provider), a name of the offering, and the input and/or output data provided to a consumer when the offering is accessed. The description may also include information about the region (e.g. the city or spatial extent) where the resources relate to, the price for accessing the resources, the license of the data provided, the access control list, etc.

Consumers discover offerings of interest on the marketplace by providing an (offering) query. The query entails a specification of the type of offerings the consumer is interested in. For example, a consumer can provide a description of the desired resources (such as type of parking information) and define the maximum price, the desired license types, the region, etc. Upon a discovery request, the marketplace identifies all matching offerings and returns them to the consumer. The consumer can then choose the offerings of interest and subscribe to those on the marketplace. Since the discovery can take place at run-time, a consumer is able to identify and subscribe to newly offered resources as they emerge. Finally, to limit the data to be transmit upon accessing an offering, a consumer can also provide a filter in the request.

3.2   Use Cases and Requirements

The high-level requirements for designing the architecture have been identified through discussion of relevant use cases and from a qualitative survey among the stakeholders from industry and research involved in the BIG IoT project. Clusters of requirements have been identified, as described in the following.

(1) **Core technology** – Given the overall goals of our work, namely to facilitate IoT ecosystem creation, and to enable resource providers to trade and monetize their IoT resources and consumers to discover and utilize them across platform and domain boundaries, we have identified the following high-level functional requirements. First, IoT platforms and services need to be able to offer and register IoT resources on a marketplace, and provide easy access to the resources via a common API. Second, applications and services shall be able to discover desired IoT resources via a marketplace and access them across heterogeneous platforms or services via a common API. Third, resource providers shall be able to monetize their assets (information and functions) via a marketplace. Fourth, resource consumers shall be able to discover new resource providers at run-time and leverage their resources immediately.

In conclusion, we identified three technological pillars that are key for the development of an IoT ecosystem: a centralized marketplace, common API(s), and a software development kit (SDK) for easy integration with the ecosystem. The API and its implementation, the SDK, need to be developed in an open source/ community process.

(2) **Developer support** – In order to grow an IoT ecosystem, it is crucial to lower the hurdle of joining the ecosystem, and thus, support developers in the process of extending their IoT platforms, services or applications. These scenarios involve developers that (a) extend their platform to support the common API and offer resources to a central marketplace, and (b) develop a service or application, which uses the common API to gain access to the marketplace to discover offerings and connect to their provider platforms or services. In this context, we identified three essential use cases. First, a developer studies the BIG IoT documentation, example

code and downloads the SDK. Second, a service/platform developer implements a service or extends an IoT platform to register a resource offering on the marketplace. Third, an application/service developer implements an application/service, which utilizes a resource offering discovered via a marketplace.

(3) **Exchange of resource offerings** – This cluster of use cases defines how (a) providers can offer their resources on a marketplace, and how (b) consumers can search for offerings and access them. The derived requirements are: first, a service/platform registers a resource offering on a marketplace; and second, a service/application discovers offerings via a marketplace and accesses them on the platforms/services. Both, the registration as well as the discovery of offerings need to be supported at run-time, in order to allow consumers to leverage emerging resources as they become available.

(4) **Charging and billing** – One of the core functionalities of an IoT ecosystem marketplace is to enable providers to monetize the access and use of their resources. Therefore, the following two requirements describe the collection of accounting data as well as the necessary functions for charging and billing. First, platform/service/application instances perform accounting of the accessed resources. Second, a marketplace offers accounting and charging information to the involved stakeholders.

(5) **Non-functional requirements** – First, the integration of existing and new IoT applications, services, and platforms with a marketplace shall be low-effort. Second, the common API and the marketplace implementations shall be highly scalable to support large-scale IoT deployments. Third, the communications and interactions among consumers, providers, and the marketplace shall be secure, as this is a crucial aspect for any IoT deployment to work.

3.3 Platform Integration Modes

For the integration of heterogeneous IoT platforms into IoT ecosystems, we have analyzed the needs and constraints of the 8 platforms involved by the BIG IoT project partners (see Sect. 1). The following challenges have been identified:

1. The implementation of the API for interaction with the marketplace, and to offer access to consumers must be low effort.
2. Platform providers that use off-the-shelf platform solutions, and thus have no access to the source code of their platform, need alternative means to integrate their platforms into an ecosystem.
3. Constrained[10] device-level platform providers need infrastructure-level support to overcome the availability and cost limitations of such platforms.

In order to address *challenge 1*, BIG IoT provides developers an SDK comprising a library for using the API. This way, a developer can extend an IoT platform

---

[10] *Constrained* in this context means that the platforms may not always be accessible (either due to energy saving reasons or with wireless coverage) and/or their backhaul connection might incur costs based on a "pay-per-use" plan (e.g. mobile phones or battery-powered sensors).

programmatically by means of an easy-to-use programming interface. While we currently focus on Java, the SDK will be provided for common programming languages and development environments.

To cope with *challenge 2*, we suggest that affected platform providers develop a gateway-service. Such a gateway-service sits between the existing platform, and the marketplace or consumer applications/services. We envision that open source gateway-service implementations will become available for common IoT platform types.

In order to deal with *challenge 3*, we support affected platform providers by releasing an open source proxy-service implementation together with an extended SDK that allows easy integration of such constrained device-level platforms with the proxy-service and the marketplace. The main functionality of the proxy-service is to store informational resources that are offered by the device-level platform and serve them to interested consumers upon request. With respect to tasks or actions that need to be processed by such device-level platforms, the proxy-service queues them until the platform connects and pulls the received tasks or actions. The response of a task or action is also proxied by such a service.

We validated the different options with all the platform providers involved in BIG IoT. The results show that 5 out of 8 platform providers are interested in the API library to extend their platform programmatically. In addition, 5 out of 8 providers indicated interest in the gateway-service based integration option. From the 2 device-level platform providers involved in the project, both confirmed interest in the proxy-service.

3.4   High-Level Design Decisions

This section draws high-level design decisions for the architecture work based on the surveyed needs of the BIG IoT platform providers and the considerations above.

1. *Focus the Marketplace Functionality on an IoT Resource Exchange.*
   The functional scope of a marketplace in an IoT ecosystem can be broad. We evaluated the following possible key functional options:
   - Resource exchange – for IoT resource providers and consumers to publish and discover their resource offerings and facilitate the resource exchange;
   - Application or service store – for IoT developers to trade their applications or services software; and
   - Hosting environment – for application, service or platform providers to host their run-time systems.

   Based on a survey among the BIG IoT partners, we have identified the resource exchange functionality as most crucial, and thus, focus the BIG IoT marketplace on this. Nevertheless, the marketplace may be extended towards other functionalities in the future.
2. *Consumers Access IoT Resources Directly on the Provider.*
   For scalability reasons and to keep IoT resources under full control of the providers we propose not to store IoT data on a marketplace, but to enable easy access to the resources directly on the provider end. This design decision has the advantage that the marketplace only requires the ability to publish and discover the resource

offerings (i.e., the descriptions of the resources), and to facilitate the direct access (e.g., through authentication of consumers and accounting support), but the actual resources remain stored and managed on the provider infrastructure.

3. *Providers and Consumers Can Participate on Multiple Marketplaces.*
   In order to avoid a marketplace lock-in, we propose to allow providers and consumers to use and interact with multiple marketplace instances at the same time. The advantage is that providers can offer their resources on multiple marketplaces, and thus, minimize the risk of integrating the API without good prospects to regain the initial investment of joining the ecosystem or running the risk of a vendor lock-in. Likewise, consumers can participate on multiple marketplaces.

4   The BIG IoT Architecture

This section describes the BIG IoT architecture as a realization of the generic concepts and requirements for IoT ecosystems presented in Sect. 3. A first implementation by the BIG IoT project is currently in progress. As shown in Fig. 2, we distinguish the following 5 core building blocks:

(1) **BIG IoT enabled Platform** – this IoT platform implements (as a *provider*) the common API, which is called the BIG IoT API, to register offerings on a BIG IoT Marketplace, and grants BIG IoT Services or Applications (as *consumers*) access to the offered resources.

(2) **BIG IoT Application** – this application software implements and uses the BIG IoT API, (as a *consumer*) to discover offerings on a BIG IoT Marketplace, and to access the resources provided by one or more BIG IoT Services or Plat-forms (as *providers*).

(3) **BIG IoT Service** – this IoT service implements and uses the BIG IoT API to register offerings on a BIG IoT Marketplace (as a *provider*) and/or to discover and access offerings provided via a BIG IoT Marketplace (as a *consumer*).

(4) **BIG IoT Marketplace** – this composite system consists of sub-components: The *Marketplace API* serves as an entry point for all communications and interactions with the marketplace; the *Identity Management Service (IdM)* which authenticates and authorizes providers and consumers; the eXchange, which allows registration and discovery of offerings using semantic technologies; the *Web Portal* for users of the Marketplace; and the *Charging Service*, which collects accounting information. The Web Portal allows the users of a marketplace (typically organizations) to register and create accounts for their developers and administration personnel who in turn can create and register new provider or consumer instances, define new offerings and queries (for supported application domains), query and subscribe to offerings of interest, and manage those conveniently via a Web browser.

(5) **BIG IoT Lib** – this is an implementation of the *BIG IoT API* that supports service and application developers. The BIG IoT Lib consists of a *Provider Lib* and a *Consumer Lib* part. It translates function calls from the respective application or service logic, or the platform code into interactions with the marketplace, or peer-services or -platforms. The Provider Lib allows a platform or service to authenticate

itself on a marketplace and to register offerings. As described in Sect. 3.1, the offering description is machine-readable and we base it on RDF [15] models. It incorporates the W3C WoT [13] Thing Description design pattern: offerings can be semantically described by integrating existing domain contexts (e.g., specific vocabularies for smart cities, smart home, or manufacturing). The Consumer Lib allows an application or service to authenticate itself on a marketplace, to discover available offerings based on semantic queries, and to subscribe to offerings of interest. The use of semantic technologies enables the *BIG IoT eXchange* to perform semantic matching even in case providers and consumers use different semantic models or formats, as long as a common meta-model defines the relations/ mapping between the different semantic models and converters for the different semantic formats are supported.

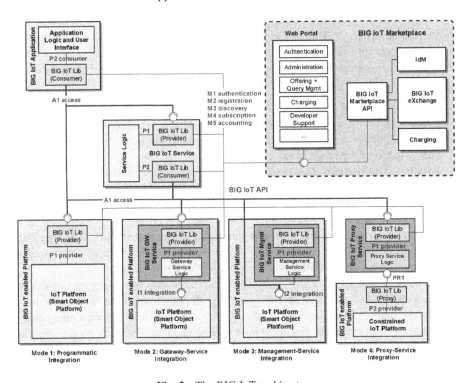

Fig. 2. The BIG IoT architecture.

4.1 Architecture Integration Modes

To Comply with the Requirements Identified in Sect. 3.3, the Architecture Supports the Following Platform Integration Modes

Mode 1: the platform developer uses the *Programming Interface P1* provided by the Provider Lib to extend an existing or new IoT platform programmatically.

Mode 2: the provider develops and operates a *BIG IoT Gateway Service*, which handles all BIG IoT related interactions and translates the relevant requests into calls supported by the existing platform (*Integration Interface I1*).

Mode 3: the provider develops and operates a *BIG IoT Management Service*, which handles the interactions with the marketplace. It integrates with the legacy platform by implementing the *Integration Interface I2*. Access to the resource offerings is provided directly by the legacy platform.

Mode 4: the provider develops and operates a *BIG IoT Proxy Service*, which handles the interactions with the marketplace and offers the *Access Interface A1*. The proxy-service acts as an "always-available" proxy on behalf of a typically constrained device-level platform.

4.2   Interfaces and Interactions

Besides the core components, Fig. 2 also depicts the relevant interfaces of the architecture. The *Programming Interfaces P1* and *P2*, provided by the BIG IoT Lib, are offered to developers to connect their components with the marketplace and other entities. For easy integration of constrained device-level platforms, a special BIG IoT Lib is provided, which allows developers to interact easily with the BIG IoT Proxy Service (via the *Programming Interface P3*).

The BIG IoT Marketplace provides five interfaces to allow interactions with its services. The *M1 interface* is used by the provider and consumer instances to authenticate themselves on the marketplace at start-up. Upon successful authentication, the Provider or Consumer Libs will obtain the required credentials for any further communication and interaction with the marketplace. The *M2 interface* is used by providers to register/deregister offerings, while the *M3 interface* is used by consumers to discover them on the marketplace. Once a registration request is received, the BIG IoT eXchange validates the offerings and stores them in a semantic database. To subscribe/unsubscribe to offerings, consumer applications use the *M4 interface*. With a subscription, a consumer indicates its intent to access the offered resources, and confirms its consent with respect to the offering's license, price, etc. Once an offering is subscribed, the eXchange provides the consumer unique credentials to access this offering. In case the offering has expired or has been updated by the provider, the eXchange revokes the subscription and indicates the cause in the response. The *M5 interface* is used by consumers and providers to send accounting information in regular time intervals to the Charging Service. Accounting types (e.g. per message) can differ between offerings, and are specified by a provider in the offering description. The interfaces M1–M4 are used in the same way by the Web Portal.

The *Access Interface A1* is the interface via which a consumer gets access to resources offered by a provider. Depending on the Provider Lib implementation, it will support different access means. All Provider Libs shall support the HTTP-based request/response access. Optionally, a Provider Lib can also support other protocols (e.g. WebSockets, MQTT) or other access paradigms (e.g. streaming).

Figure 3 describes the discovery (M3) and subscription (M4) to offerings on the marketplace in more detail. Once a query has been created by a developer via the Web

UI, we distinguish between two modes: static and dynamic. In static mode, the developer or administrator of a consumer application or service selects and subscribes to the offerings of interest manually, via the Web portal. In dynamic mode, queries can be refined by the application or service logic programmatically, e.g. in order to consider information that is only available at run-time (e.g. location) and the subscriptions to offerings is automated based on consumer-defined policies. The dynamic mode is needed in case an application or service is designed to discover and integrate new data sources at run-time, e.g. in order to incorporate emerging offering providers automatically.

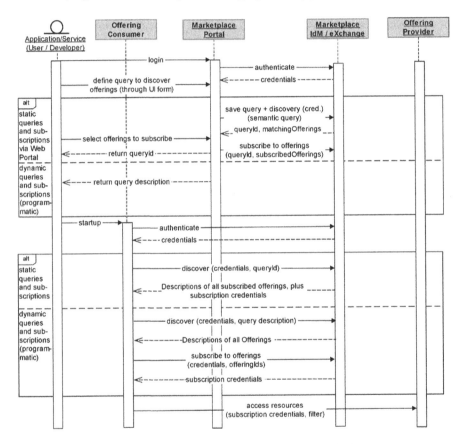

Fig. 3. Sequence of offering discovery and subscription.

5 Proof-of-Concept Implementation and Demonstrator

This section presents a proof-of-concept implementation of the BIG IoT architecture components. In an end-to-end scenario, the practicability of the BIG IoT architecture, including the marketplace and the API for an interoperable IoT ecosystem and the feasibility of the approach is demonstrated.

The overall goal of the developed architecture is to ease the interoperation of IoT platforms, services and applications despite technological and organizational boundaries. This scenario showcases that the run-time discovery and integration of IoT resources provided by heterogeneous platforms and various organizations becomes possible through the developed API and the marketplace. Although this scenario incorporates platforms from the smart city domain, the BIG IoT components and interfaces can be utilized in other domains as well. The key components are (1) the eXchange backend and the Web Portal of the BIG IoT Marketplace, (2) a demo Web application, (3) the cloud-level OpenIoT platform offering parking space data, and (4) the device-level Wubby platform offering air quality data. Figure 4 shows the key components as well as the interfaces and connections between those components. By default, the two platforms shown at the bottom offer their own proprietary interfaces. To integrate these platforms into the BIG IoT ecosystem, two gateway-services (shown in blue) are implemented according to integration mode 2 (Sect. 4.1). Those gateway-services implement the adaptation of the proprietary platform interfaces to the BIG IoT API. This adaptation is facilitated through usage of the BIG IoT Lib (shown in yellow). This library offers the access interface for consumers to access the resources, and can be used to interact with the marketplace as a client. The library is also used by the demo Web application (shown in green). Using the BIG IoT Lib, simple method calls in the particular programming languages (here: Java), makes it easy to discover the relevant resource providers and to utilize the access interface of the heterogeneous platform.

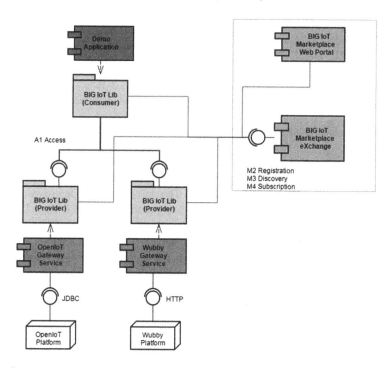

Fig. 4. Components of the proof-of-concept demonstrator implementation. (Color figure online)

The sequence of interactions in this demonstration is illustrated in the following. An application developer is implementing a Web application (Fig. 6) that is supposed to visualize available parking spaces in smart cities. First, the developer visits the marketplace Web Portal and fills out the UI form accordingly to search for available resource offerings based on a semantic type that is of interest to her. Figure 5 shows the screenshot of the prototypical implementation of the marketplace Web Portal. We assume that at this point, only a few parking information offerings are found for this category. Nevertheless, the user is presented with a *Query ID* that is associated with the respective parameters. The user utilizes the returned *Query ID* and places it in her application code in order to allow her Web application to perform regular discovery requests for the offerings interested on the marketplace. Running the application triggers the discovery request based on the *Query ID*, however, only few parking spaces are shown on the map, as not many offerings of type "parking" are registered or active.

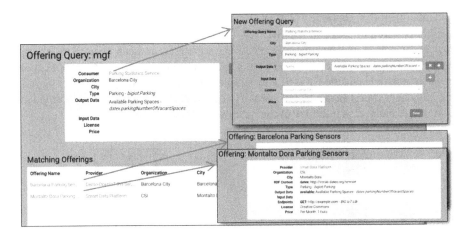

Fig. 5. Screenshot of the marketplace portal UI to create and view queries.

In a next step, a new user (platform provider) visits the marketplace portal to create an offering called "Barcelona Parking Sensors" and tags it with the same semantic type. After this creation, the offering is still inactive. The portal presents the provider with an *Offering ID*. This ID is used by the provider as a parameter in the OpenIoT gateway service. Once the provider starts the gateway service, it automatically registers the offering on the marketplace using the created *Offering ID* and marks it then as active.

Coming back to the Web application, which makes periodic discovery requests based on the defined Query ID, it now finds the new offering (of the desired semantic type) and automatically accesses and integrates the data in the application. As a result, the application visualizes the newly found parking spaces as markers at their specific locations (Fig. 6).

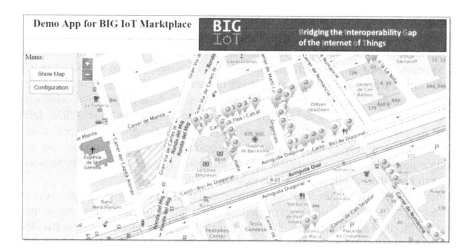

Fig. 6. Screenshot of the Web application to view the available parking spots.

Once the application receives a new offering from the marketplace, it checks all relevant information (e.g., price for accessing the offering, or license agreement) whether they meet the user's requirements. Then, the application subscribes to the matching offering, and eventually accesses the offering to retrieve the parking information. The access to the parking information on the provider platform is enabled by using a direct access interface provided by the BIG IoT Provider Lib. When the Web application calls the access method (provided by the BIG IoT Consumer Lib), the OpenIoT Gateway Service translates the requests for the parking information into a proprietary call to the OpenIoT platform and returns the data to the Web application.

The validation of the BIG IoT architecture in other use case scenarios and across domains is ongoing work in the project, aiming to address the five interoperability patterns identified and presented in [6].

6   Conclusions and Outlook

Grounded in our vision of interoperable IoT ecosystems [6], we define, in this article, generic concepts for IoT ecosystem architectures, such as marketplaces, offerings, providers, consumers. Based on this core terminology, we present guiding use cases and requirements for the architecture, which were derived from surveys among the industrial and research partners of the BIG IoT project. We realize those generic considerations in the concrete BIG IoT architecture, by describing the key building blocks, their interfaces, and interactions. Finally, we present a first proof-of-concept implementation and demonstrator in order to illustrate the core architectural concepts, their feasibility, and the advantages of the architecture. While this scenario incorporates IoT platforms from the smart city domain, the BIG IoT components and interfaces are likewise applicable in other domains.

The demonstrator shows that the defined architecture is capable of: (1) solving the discovery challenge of available IoT resources for application and service providers, despite the fact that resources are collected and stored across heterogeneous platforms and systems, across large geographic spaces, and by a multitude of stakeholders and organizations, who are mostly not even aware of each other; (2) bridging the interoperability gap among heterogeneous IoT applications, services and platforms, which are using various standards and technologies, and operate on different scales (cloud-level vs. device-level platforms); and (3) addressing the evolvability problem of applications and services, who rely mostly on manual integration of continuously emerging IoT resource providers (e.g., new data sources), and thus, require growing development efforts to keep their applications or services up-to-date.

First, the BIG IoT architecture, with its marketplace, overcomes those challenges by introducing "places" for resource providers and consumers to meet and exchange their resource offerings and demands, and discover each other. Second, based on the BIG IoT API, the heterogeneous platforms and systems involved are able to access and exchange resources using standard protocols and frameworks. Finally, since the BIG IoT architecture supports the discovery of providers and their resources as well as the access to the resources at run-time, IoT applications and services are now able to integrate automatically emerging resource providers at run-time.

Key enablers for addressing the discovery challenge are semantic technologies. They facilitate the matching of resource offerings and queries across heterogeneous systems and diverse stakeholders, and also help to overcome the interoperability challenge. In the future, semantic vocabularies for specific application domains need to be established. This is needed in order to enable semantic matchmaking for IoT offering discovery on the marketplace. The BIG IoT project aims at using and extending existing and proven vocabularies, such as schema.org.

The detailed specification of the BIG IoT API, and in particular the use of semantic technologies to describe resource offerings, queries, the resources themselves, as well as the detailed specification of the BIG IoT Marketplace architecture, including the eXchange and the use of semantic databases, is ongoing work. To ground these specifications in public standards, we are actively contributing to the W3C Web of Things group and will continue doing so in the future.

Acknowledgments. This work is financially supported by the project "Bridging the Interoperability Gap" (BIG IoT) funded by the European Commission's Horizon 2020 research and innovation program under grant agreement No. 688038.

References

1. Gershenfeld, N., Krikorian, R., Cohen, D.: The internet of things. Sci. Am. **291**, 76–81 (2004)
2. Manyika, J., Chui, M., Bisson, P., Woetzel, J., Dobbs, R., Bughin, J., Aharon, D.: The internet of things: mapping the value beyond the hype. McKinsey Global Institute (2015)
3. Alliance, O.M.: Lightweight Machine to Machine Technical Specification, Candidate (2015)
4. Bröring, A., Echterhoff, J., Jirka, S., Simonis, I., Everding, T., Stasch, C., Liang, S., Lemmens, R.: New generation sensor web enablement. Sensors **11**, 2652–2699 (2011)

5. Swetina, J., Lu, G., Jacobs, P., Ennesser, F., Song, J.: Toward a standardized common M2M service layer platform: introduction to oneM2M. IEEE Wirel. Commun. **21**, 20–26 (2014)
6. Bröring, A., Schmid, S., Schindhelm, C.-K., Khelil, A., Kaebisch, S., Kramer, D., Le Phuoc, D., Mitic, J., Anicic, D., Teniente, E.: Enabling IoT Ecosystems through Platform Interoperability. IEEE Softw. (forthcoming) (2017)
7. Soldatos, J., et al.: OpenIoT: open source internet-of-things in the cloud. In: Podnar Žarko, I., Pripužić, K., Serrano, M. (eds.) Interoperability and open-source solutions for the internet of things. LNCS, vol. 9001, pp. 13–25. Springer, Cham (2015). doi: 10.1007/978-3-319-16546-2_3
8. Bassi, A., Bauer, M., Fiedler, M., Kramp, T., Van Kranenburg, R., Lange, S., Meissner, S.: Enabling things to talk, Designing IoT solutions with the IoT Architectural Reference Model. Springer, Heidelberg (2013)
9. Ramparany, F., Marquez, F.G., Soriano, J., Elsaleh, T.: Handling smart environment devices, data and services at the semantic level with the FI-WARE core platform. In: IEEE International Conference on Big Data (2014)
10. Soldatos, J., Aikaterini, R., Kaldis, J.: VITAL - Virtualization Architecture and Technical Specifications, vol. D2.3, European Commission - FP7 (2015)
11. Barnaghi, P., Tönjes, R., Höller, J., Hauswirth, M., Sheth, A., Anantharam, P.: Citypulse: real-time iot stream processing and large-scale data analytics for smart city applications. In: Europen Semantic Web Conference (ESWC) 2014 (2014)
12. Houyou, A.M., Huth, H.-P., Kloukinas, C., Trsek, H., Rotondi, D.: Agile manufacturing: general challenges and an IoT@Work perspective. In: 17th IEEE International Conference on Emerging Technologies & Factory Automation (ETFA 2012) (2012)
13. World Wide Web Consortium (W3C), Web of Things (WoT). https://www.w3.org/WoT/
14. Mineraud, J., Mazhelis, O., Su, X., Tarkoma, S.: A gap analysis of Internet-of-Things platforms. Comput. Commun. **89–90**, 5–16 (2016)
15. Klyne, G., Carrol, J.J.: Resource Description Framework (RDF): Concepts and Abstract Syntax, W3C Recommendation, W3C (2004)
16. Davies, J.: Hypercat: resource discovery on the internet of things. IEEE IoT Newsl., 12 January 2016. http://iot.ieee.org/newsletter/january-2016/hypercat-resource-discovery-on-the-internet-of-things.html

A Sensor Observation Service Extension for Internet of Things

Argyrios Samourkasidis and Ioannis N. Athanasiadis(✉)

Information Technology Group, Wageningen University,
Hollandseweg 1, 6706 KN Wageningen, The Netherlands
{argyrios.samourkasidis,ioannis.athanasiadis}@wur.nl

Abstract. This work contributes towards extending OGC Sensor Observation Service to become ready for Internet of Things, i.e. can be employed by devices with limited capabilities or opportunistic internet connection. We present an extension based on progressive data transmission, which by-design facilitates selective data harvesting and disruption-tolerant communication. The extension economizes resources, while respects the SOS specification requirement that the client should have no a-priori knowledge of the server capabilities. Empirical experiments in two case studies demonstrate that the extension adds little overhead and may lead to significant performance improvements in certain cases, as for irregular timeseries. Also, the proposed extension is not invasive and backwards compatible with legacy clients.

Keywords: Open Geospatial Consortium · Sensor Observation Service · Internet of Things · Syntactic interoperability · SOS 2.0 · Sensor Web · Progressive transmission · Pagination · Timeseries data

1 Introduction

Internet of the Things (IoT) is a dynamic, open, participatory ecosystem of decentralized and collaborative devices. Recent technological advances resulted in a plethora of low-cost devices with extended capabilities compared to traditional sensors. New generation of devices are miniaturized and empowered with storage, processing and networking capacity. They are essentially transformed into **smart nodes**, that operate autonomously, may offer added value services [31], and collaborate with each other in the cloud [4]. Smart nodes could offer capture, storage and dissemination services of sensory information in a single device [36]. IoT devices are also instrumental to the proliferation of new data sources [14], sharing of information [15], and contribute to the *big data* movement. Internet of Things advances the vision of Sensor Web, *an infrastructure which enables interoperable usage of sensor resources* [8]. In the IoT era, Sensor Web is challenged to offer services that are interoperable, but at the same time perform efficiently with **less resources**, saving processing power and network bandwidth.

© Springer International Publishing AG 2017
I. Podnar Žarko et al. (Eds.): InterOSS-IoT 2016, LNCS 10218, pp. 56–71, 2017.
DOI: 10.1007/978-3-319-56877-5_4

Interoperable data interchange for sensor data has been driven by the Open Geospatial Consortium (OGC). OGC introduced service interfaces and information models within Sensor Web Enablement (SWE), which is founded on machine-to-machine communication [5,7]. Service interfaces, as the Sensor Observation Service, Web Feature Service, Web Coverage Service, SensorThings provide interoperable means for geospatial information discovery and retrieval. Sensor Observation Service (SOS) [24,25] is an OGC service interface, which promotes interoperable sensor-borne data exchange, operates as a web service, and supports for syntactic and semantic interoperability.

In the IoT era, architectural paradigms and technologies need to respect the limited capabilities of devices. The SWE 2.0 has been established with technologies as the Simple Object Access Protocol (SOAP) and XML-based information models, which are considered to add substantial overhead - a critical issue for IoT devices. On the other hand, Representational State Transfer (REST) and JSON-based information models seem to provide services which excel over SOAP and XML, in terms of power consumption and performance [22]. Beyond these technical limitations, there are certain *design* choices that preclude SOS as an appropriate IoT outlet.

In this paper, we investigate current SOS design and propose an extension. In Sect. 2, we present related work, how SOS operates and challenges identified in the literature. In Sect. 3, we identify SOS design shortcomings from an IoT perspective, and introduce a pagination technique in order to promote selective data harvesting, enable seamless data integration and facilitate machine-to-machine interoperability. Section 4 presents an implementation and details the two case studies, which were designed to test the efficiency of the extension, along with experimental results. Section 5 provides with a discussion about our findings and contributions, concludes the research and lays the groundwork for future work.

2   Related Work

2.1   Service Orientation and Interoperability in Sensor networks

Service-Oriented Architecture (SOA) is an architectural paradigm founded on self-describing, self-contained services. Key concept in SOA is that services may be developed, maintained and served by different entities, and can subsequently be combined and produce composite applications. SOA has been instrumental for highly interoperable systems, as services are platform and language independent [30].

In the frame of interoperable data interchange, OGC introduced Sensor Web Enablement (SWE), which follows the SOA architectural paradigm. Standards developed within SWE provide means for the discovery and retrieval of sensor observations. SWE contributes towards the vision of Sensor Web, where web-accessible sensor networks and archived sensor observations can be discovered and accessed using standard protocols and application program interfaces (APIs) [5]. They are realized through *web services*, i.e. services "identified by a URI, whose service description and transport utilize open Internet standards" [30]. Communication between service interfaces and other services or clients is achieved through Simple Object Access Protocol (SOAP), which builds on existing communication layers (i.e. HTTP) [10]. SWE is a very important infrastructure [8] as it offers interoperable protocols for advertising, disseminating and requesting data among heterogeneous sensor systems and devices.

2.2 The Sensor Observation Service

Sensor Observation Service (SOS) is an OGC service interface specification for accessing sensor observations, which acts as "the intermediary between a client and an observation repository" [5]. SOS interface enables clients to request, filter and retrieve observations, and metadata about repositories and sensors.

SOS comes with a *core* set of services, and *extensions* that enrich it with extra functionality, or *profiles* for domain-specific behavior. The current 2.0 specification [24] defines three *core* operations:

a. service discovery (`GetCapabilities`),
b. sensors metadata retrieval (`DescribeSensor`), and
c. observations retrieval (`GetObservation`).

There are several extensions and profiles available, but their description falls outside the scope of this paper. As an indicative example for the reader, the *transactional extension* provides with services to register new sensors and add new observations.

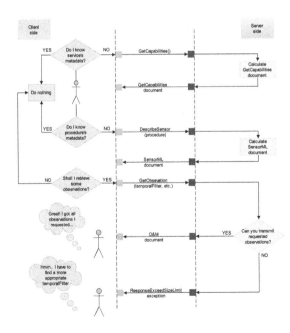

Fig. 1. A typical observation retrieval workflow using SOS

SOS is a pull-based service interface and is intended for machine-to-machine communication. The protocol prescribes a communication between a client and a server, both can be considered to be software agents. The client submits a request and the server answers with a response, typically in the form of XML document. Responses are encoded in appropriate SWE related XML schemas as Observation & Measurements [23], or SensorML [27]. A typical observation retrieval workflow using SOS is depicted in Fig. 1. First, the client inquires the server for its capabilities. Then, it may ask for descriptions on certain sensors, and finally requests for observations from one or more sensors. A typical Get-Observation request includes temporal and/or spatial boundaries.

When SOS server encounters an error while performing a GetObservation operation, it returns an exception. For example, if client asks for wrong values of arguments an InvalidParameterValue exception is rendered. In the current SOS 2.0 interface standard [24] there is also a type of exception for the cases that the response exceeds a **size limit**. We will investigate this further below.

2.3   Challenges in Sharing Sensor Observations in IoT

Internet of Things consists of *smart nodes* equipped with sensors and network connectivity, able to interact with their environment and share information. *Smart nodes* are entitled with specific characteristics:

a. restrained capabilities (in terms of energy and processing power),
b. opportunistic Internet connection, and
c. heterogeneity in resulting data formats and communication protocols [3].

Key challenges towards the IoT realization include energy efficiency, integration of service technologies and security/privacy [21]. Also, thematic and spatial concerns of deployed IoT systems pose great challenges in spatiotemporal aggregation of disperse observation datasets.

As regards with **heterogeneous sensor integration**, previous studies have been conducted towards various directions. A virtual integration framework for heterogeneous meteorological and oceanographic data sources is demonstrated in [33]. A *SOS profile* to facilitate multi-agency sensor data integration was reported in [1,19]. Fredericks et al. argue in [12] that quality metadata should also be transmitted through SWE services, in order the realization of automatic data integration to be achieved.

Integration of spatially diverse sensor timeseries utilizing OGC standards concerned Horita et al. in [16]. They developed a spatial decision support system for flood risk management, associating Volunteered Geographic Information (VGI) and measured data derived from Wireless Sensor Networks (WSNs). Data acquisition, integration and dissemination is orchestrated by a SOS instance.

Only recently, OGC introduced *SensorThings API* to facilitate "the interconnection of IoT devices, data, and applications over the Web" [28]. In contrast with other OGC standards, SensorThings API adopts the REST paradigm and utilizes JSON-based information models. SensorThings API defines HTTP requests to facilitate observations' retrieval, as well tasking of sensors and actuators.

Using parameters to regulate response size to requests within OGC-related standards, was a topic of interest for [26,28,29]. Lengthy responses to `Get-Observation` requests have been identified as a potential danger to both SOS server and clients [26]. In the same work, it has been indicated that beyond the `ResponseExceedSizeLimit` exception, other certain limitations as regards with the number of returned observation should be concerned and imposed. The WFS interface standard [29] and the SensorThings API offer a paging implementation, that allows the client to limit the number of features included in a response by using two optional arguments (*count*, *startindex* for WFS, and *top*, *skip* for SensorThings API).

Last but not least, several researchers investigated the suitability of limited bandwidth, energy, and processing power devices to host a SOS server. These have mainly concentrated on (a) adoption of lightweight architectural paradigms (e.g. REST instead of SOAP [17,35,39]), and (b) evaluation of SOS lightweight implementations [18,32]. We have also deployed SOS over a Raspberry Pi to exploit the potential of low-cost embedded devices [36].

In this work we concentrate on the SOS service interface design and evaluate the efficiency of communication between client and server.

3   Methods

3.1   SOS Service Interface Design Issues

According to SOS specification, clients are not allowed to know sensor observations' frequency. The server advertises the boundaries of the information it holds, but not the resolution. Any client is not possible to infer the sensor temporal or spatial resolution, based on their communication with the server. This requirement is that the client has access with **no a-priori knowledge** [25]. While this enforces reusability and generality of the service interface, it may lead to excessive data requests, which may result to server overload, or even Denial of Service attacks.

Excessive data transmission has been identified as an issue for `GetObservation` requests. In the first specification of SOS, there was not imposed any limitation, regarding the maximum number of observations which could be transmitted. For the server, the only viable response to of a `GetObservation` request was to return a set of observations. The server had no way to refuse to respond, in cases where the client was asking for an excessive amount of data, it was busy, or any other reason.

To illustrate the above shortfall we will consider a service offered by National Oceanic and Atmospheric Administration (NOAA) [9]. NOAA's Center for Operational Oceanographic Products and Services (CO-OPS) offers openly a variety of sensor observations using SOS. In this implementation, if a client requests observations for a time range which exceeds 31 days, the server responds with an exception, rejecting the parameter value:

```
<Exception exceptionCode="InvalidParameterValue"
    locator="eventTime">
        <ExceptionText>
            Max 31 days of data can be requested.
            62.0 days were requested.
        </ExceptionText>
</Exception>
```

Note that the exception rejects the parameter value, disclosing in a non machine interoperable message of the size limits for this request.

In the future work section of SOS 1.0 specification [25] it was acknowledged that: *"The density of requests and offerings must be addressed,... so that large data volumes are not transmitted unnecessarily due to a lack of information about service offerings."*. Indeed, that was addressed in SOS 2.0 by introducing an *exception* to manage excessive data requests, while taking into consideration the *no a-priori knowledge* requirement [5]. The `ResponseExceedSizeLimit` exception functionality resembles the response of NOAA server above, but with pertinent semantics to the exception thrown: The server is able to inform the client that the *"requested result set exceeds the response size limit of the service and thus cannot be delivered"* [24]. Both server and client applications are protected from extremely big response sizes, and the *no a-priori knowledge* requirement is respected.

The `ResponseExceedSizeLimit` exception of SOS 2.0 is a significant improvement compared to SOS 1.0, as it allows the server to respond to a request with an exception than with actual data. Note that, the response size limit should not be considered a fixed parameter. It could change when there is high traffic, or service maintenance. In those conditions, the server should be allowed to not to respond to requests that would under normal conditions.

However, the main limiting factor to this design is that clients have no insights regarding the carrying capacity of the server, or (equivalently) the density of an offering. Due to the *no a-priori knowledge* requirement, clients cannot infer how to narrow down their requests so that server responds.

We identify two cases here. First case is when the server publishes regular sensor observations. Under this category fall most long-term, permanent sensor infrastructures. In this case, clients could implement heuristic techniques to discover the response size limit (assuming that it is constant).

In the second case, observation streams are irregular. This may happen if the sensor sampling frequency varies, or sensors move. For example, consider sensors operating in energy restrained environments and adopt opportunistic sensing techniques, or event-based sensing [2]. Volunteered Geographic Information Systems which enable individuals [11,13] or cars [6] as data providers, fall in the same case. In these situations, it is impossible for the client to make any kind of estimate on the response size, and devise a strategy to reduce accordingly the spatiotemporal boundaries of their query.

Responding with an exception to voluminous requests could be tolerated in fixed sensor networks (case one above). However, it hinders SOS applicability in resource-constrained environments. As clients are neither aware of the response size limits, nor how to restrict their queries, the SOS communication protocol underperforms: It wastes both processing power and network bandwidth as it is engaged in more request/response cycles. This, ultimately results in bigger response times. Such drawbacks are incompatible with the Internet of Things needs. This problem could be addressed by introducing a progressive data transmission technique described below.

3.2 The Resumption Token Technique and Open Archives Initiative

The notion of selective data retrieval was introduced in Lagoze and Van de Sompel [20]. Utilizing a *resumption token*, large and resource-demanding data transactions are fragmented into several requests/responses. The client submits a request and the server responds with a part of the result and a *resumption token*. Then the client (harvester) can use this *resumption token* in follow up requests to get the following part of its initial request. Gradually, by consecutive requests the client retrieves the all the partial answers to its initial request. This mechanism enables the server to handle with requests that have large responses, with respect to available bandwidth and/or processing power.

3.3 A Pagination Extension for SOS

SOS service interface can address IoT needs by introducing progressive data transmission. We extend the current SOS service interface with a *resumption token* parameter in the GetObservation requests. By fragmenting requests into many sequential ones, we transform SOS into a **disruption-tolerant** service interface, as clients are enabled to ask for specific observation subsets. Observations are divided and loosely packed into *pages* of certain size. The number of observations contained in a *page* (i.e. chunk of subsequent observations) is determined by the SOS server.

The observation retrieval workflow according to the proposed design is depicted in Fig. 2. The client asks for a set of observations with a GetObservation request. The server processes the request, and always responds with an O&M document. If the response exceeds the carrying capacity of the server, results will be organized in subsets (called pages), and the server response will include an additional element, called next which will point to the URL of the next page of results. The next page URL is the same as the original request, but contains an extra parameter called page, which has the role of the resumption token. The page parameter is optional: when a client request does not contain a page argument, the server responds with the first *page* of the request. The last page of the parts contains no next page element to notify the client of the end of the transmission.

In the simplest case, server carrying capacity could be an arbitrary, fixed threshold, similar to the *request size limit* of the SOS 2.0 exception. Of course, the server carrying capacity may dynamically vary according to result set properties, or server resources, enabling network load balancing, efficient use of energy, etc.

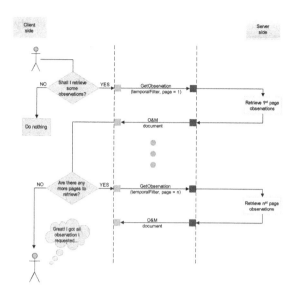

Fig. 2. A typical *paginated* observation retrieval workflow

It could even change during the transmission, as the total number of pages is not disclosed to the client. The **page** resumption token could be constructed incrementally as page number in case the server has a fixed carrying capacity, and data do not change. In case of varying page size, the **page** parameter can take unique pseudo-random integer values. In case where data changed during the communication, or any other reason, the next **page** token could be revoked by the service provider.

3.4 Expected (by Design) Benefits

The paginated protocol proposed here is beneficial for both server and client **efficiency** and **performance**. The communication protocol does not waste resources to respond with exceptions, as all requests result to responses that carry observations. This saves processing power and communication bandwidth in both client and the server.

Another attribute of the design we propose is its **non-invasive** nature. Given the *page* parameter is optional, current SOS clients can seamlessly submit **Get-Observation** requests and retrieve observations, as long as the SOS server carrying capacity is not exceeded. This means that existing SOS 1.0 or 2.0 server infrastructures could switch to a paginated implementation, and as long as they do not change their size limit threshold, existing clients would continue to operate without disruption. In the rest cases, a *page-parser* method should be implemented and incorporated in legacy clients. This method would parse a **GetObservation** response document to determine the URL of the next **GetObservation** request. On server side, the *pagination* extension could be easily applied on top of existing implementations.

4   Demonstration and Implementation

4.1   Setup

The SOS pagination extension introduced above comes with design advantages discussed in the previous section. There are also performance improvements that we experimentally evaluated by setting up two case studies. Without loss of generality, we assume not movable sensors that hold timeseries information. In case study one, the server holds a regular timeseries dataset, while in the second case study an irregular one. For both cases, we compared the SOS pagination extension (**SOS-p**) service interface against *SOS 2.0*.

The *SOS-p* server is queried by a corresponding client (*PAC*), that is able to handle **page** resumption tokens. For *SOS 2.0* server, we considered two clients: one that is not aware of *SOS 2.0* carrying capacity and finds it by employing a divide-and-conquer algorithm (*DAC*); and one that has this a-priori knowledge (*LEC*).

The three clients are in detail as follows:

Divide and Conquer client (DAC): *DAC* submits `GetObservation` requests according to *SOS 2.0* specification. When the server responds with a `Response-ExceedSizeLimit`exception, *DAC* halves the time window and submits a new query. When *DAC* finds a time window for which the server responds with no exception, it continues asking for observations with of this duration size in the temporal filter, until it has received all the data corresponding to the original request.

Leaky client (LEC): *LEC* knows the server *carrying capacity* and arranges the *temporal filter* of its request, so that there are no exceptions. While this is against the *no a-priori knowledge* requirement, it corresponds to the most favorable situation for the existing *SOS 2.0* protocol. *LEC* submits `GetObser-vation` requests to *SOS 2.0* only for case study 1.

Pagination-aware client (PAC): *PAC* client submits `GetObservation` requests according to *SOS-p*, i.e. it is capable of processing the page resumption token. In its first `GetObservation` request asks for the first page, and then processes the response for the next **page** it will ask for. If the `GetObservation` response document does not contain a next *page* tag, it means that all requested observations were transmitted.

4.2   Implementation and Synthetic Datasets

This study makes use of the AiRCHIVE SOS server implemented in Python [36]. Clients were also implemented in Python. Queries to SOS server were submitted as *HTTP GET* requests via Python Requests module [34]. Response times for each case study were facilitated using the Python Time module [38]. All experiments were carried out on a Intel Core i5 4 Mac with a 2,4 GHz and 16.0 GB of memory (1600 MHz DDR3), running OS X El Capitan (Version 10.11.1). SOS server and SOS client instances operated on the same physical machine.

In both case studies, a dataset of 15,000 observations was artificially generated. In case study one, we assumed that measurements are sensed in constant intervals of 10 s. In case study two, observations were timed with a *inconstant* frequency. Observation time interval varies from 10 to 3000 s, distributed uniformly. Timestamps were generated with the Python Random Number generator module, using Mersenne Twister [37]. Both timeseries were stored in two SQLite databases and made available to the servers.

4.3   Experimental Setup and Metrics

As limited bandwidth and processing power are key elements of IoT systems, we set up accordingly our experiments. The **carrying capacity** of the servers was defined to be 15 observations. This arbitrary threshold was chosen so that there will be significant traffic of SOS requests. *SOS 2.0* server would render a `ResponseExceedSizeLimit` exception if the result set would include more than 15 observations. *SOS-p* server organizes its responses in pages of 15 observations per page.

Clients were configured to request for observations for time intervals that result to 1 000, 2 000, 4 000, 8 000 or 15 000 observations (*response length*). Experiments have been repeated 10 times for all clients and both case studies.

For all experiments, we recorded two metrics:

a. the **response time** is the total time passed until the client has received the total amount of data requested. Measured in *seconds*.
b. **transfer volume** is the total size of all response documents received by the client until the whole response has been received. It is measured in *MB*.

Response times are averaged across the 10 repetitions, while transfer volume is the same for each repetition.

For the cases of **SOS 2.0** implementation, in the *average response time* and *transfer volume*, time spent and resulted size of exceptions are also included.

4.4   Experimental Results

Tables 1 and 2 summarize the results for both case studies and all clients. The *response time* is reported as average and standard deviation of ten repetitions.

For case study 1, best results are achieved, as expected, by the client that is aware of the server carrying capacity (*LEC*), but violates the *no a-priori knowledge* requirement. The divide-and-conquer client (*DAC*) in *SOS 2.0* adds an overhead to the transmission, as it needs to search for a working time interval. Its performance is affected mostly of how close the time interval found is to the servers carrying capacity. The response time was significantly increased in

Table 1. Experimental results for the **regular** timeseries for all three clients. Average response times and standard deviation across ten requests are reported. Total volume of the data transmitted, number of requests, and number of exceptions for DAC.

Query length	PAC			LEC			DAC		
	Resp. time (std) [s]	Vol [MB]	Reqs	Resp. time (std) [s]	Vol [MB]	Resp. time (std) [s]	Vol [MB]	Exceptions
1000	1.34 (±0.049)	0.59	67	1.29 (±0.013)	0.58	2.36 (±0.02)	0.63	7
2000	2.68 (±0.012)	1.2	134	2.57 (±0.011)	1.2	4.64 (±0.05)	1.3	8
4000	5.53 (±0.083)	2.4	267	5.22 (±0.017)	2.3	9.27 (±0.03)	2.5	9
8000	11.93 (±0.031)	4.7	534	10.31 (±0.034)	4.7	18.31 (±0.06)	5.0	10
15000	24.77 (±0.739)	8.9	1000	19.05 (±0.043)	8.7	21.33 (±0.06)	8.8	10

our experiments in Table 1. In the contrary, the performance of SOS-p and the paginated client PAC is very close to the server carrying capacity, without any breach of the *no a-priori knowledge* requirement. Experimental results in Table 1 illustrate overheads less than 5% in response time for up to few hundreds of pages, while for bigger numbers of requests overheads in time may end up to 30% in response time. This is attributed to the efficiency of the pagination implementation and is a well-known limitation among the database community. In the future, we will investigate other database options that can improve this further.

For case study 2, irregular timeseries are served therefor there is no notion of leaking the prior knowledge of the server carrying capacity. Here the paginated SOS-p excels over SOS *2.0*, as presented in Table 2. SOS-p and PAC are faster than SOS *2.0* by more than 60% on average on every GetObservation request. Also, note that number of requests has been roughly doubled, which results to a noticeable difference in the amount data transmitted. This is to be expected, as the *divide and conquer* strategy may end up finding a query window that is far from what can be actually served. There could be other search algorithms employed for improving DAC performance. However, it is made clear from this

Table 2. Experimental results for the **irregular** timeseries. For PAC and DAC clients reports average response times and standard deviation across ten requests. Total volume of the data transmitted, number of requests and number of exceptions for DAC.

Query length	PAC			DAC		
	Resp. Time (std) [s]	Vol [MB]	Reqs	Resp. Time (std) [s]	Vol [MB]	Exceptions
1000	1.35 (±0.02)	0.59	67	2.40 (±0.03)	0.63	7
2000	2.75 (±0.05)	1.2	134	4.71 (±0.05)	1.3	8
4000	5.66 (±0.07)	2.4	267	9.28 (±0.06)	2.5	9
8000	11.97 (±0.08)	4.7	534	18.34 (±0.03)	5.0	10
15000	24.62 (±0.11)	8.9	1000	36.81 (±0.88)	9.5	11

experiment, that the paginated protocol guarantees **by design** that the optimal number of measurements is included in each response. *SOS-p* entrusts the burden of coordinating the observation boundaries to the server, which knows its limits, than having the client wasting resources with requests of suboptimal lengths. The improved performance ensures that there is no waste of resources on both the client and the server side.

5   Discussion and Conclusions

This work contributes towards improving OGC SOS protocol to become IoT ready. Drafting on top of IoT requirements as efficient resource utilization and opportunistic Internet connection, and taking into consideration response size to `GetObservation` requests requirements set in [26], we designed a SOS extension, which implements a *pagination* mechanism.

There is a fundamental difference between our design and the paging mechanism introduced in OGC WFS [29]. WFS paging design contradicts with the rationale of SOS `ResponseExceedSizeLimit` exception, that is to enable the SOS server to manage efficiently its resources. Conversely, it allows clients to select the number of returned observations, which is a feature that can only facilitate specific applications (e.g. Graphical User Interfaces which can visualize a certain number of observations). In the contrary, the solution proposed in this work follows the Open Archives Initiative design pattern, and the decision on the *page size* remains with the server, not the client. As we demonstrated above, this is a necessary condition for the server in the IoT era, as it allows for parsimonious use of resources, and protection from queries resulting with very big results.

Pagination introduces the notion of **progressive transmission**, which fits for purpose with timeseries data sequential nature, but is also suitable for any kind of spatiotemporal requests. It adds **disruption-tolerance** as an additional SOS feature, since a client can request for and retrieve a specific page. This is very useful when big datasets are to be transmitted or when the Internet connection is poor. Our design enables a SOS server to exploit its resources to the maximum, as computational power and network bandwidth are spent for yielding results, not for handling exceptions. Thus, the paginated extension enables **by-design** SOS for devices with restrained capabilities, where resources are economized in sharing interoperable knowledge.

Whilst our suggested design entails new improvements to the existing *SOS 2.0*, its importance is highlighted by its **non-invasive** nature. Backwards compatible design is achieved through the *optional* page parameter, since all requested data could be included in one *page*. This way, current *SOS 2.0* clients could operate without further modifications with *SOS-p* extended servers, if the server always responds with the whole data requested.

Evaluating the *SOS-p* extension against specific metrics, we validated improvements by experiments. Those improvements are mainly concerned with efficiency. Lower `GetObservation` requests completion times contribute towards

IoT devices energy conservation, since computational resources are occupied for less time, and thus more clients can be served simultaneously. In addition to that, when carrying capacity is not known to the client, the *SOS 2.0* protocol is under-operating, as possibly transmits less observations in each request. This results to more request-response transactions, with overheads in data volume and duration time.

The pagination extension introduced here offers a remedy to *SOS 2.0* shortfalls in handling exceptions, by providing a machine interoperable solution. It also fills-in the SOS missing piece, that is to *"allow a client to determine the density of an offering"* [25]. Instead of that, it delegates to the server to drive protocol.

Advancements discussed so far lay the groundwork for future work. Firstly, our intention to use *pagination* was exploratory, thus there is room for further improvements in the implementation to further improve performance. One direction for improvement is the adoption of a caching mechanism. *Pagination* is a good candidate for caching techniques, since requests are incremental and queries are submitted sequentially. With the design introduced here, the client reveals its intentions to the server, by asking the whole spatiotemporal boundaries of interest. If the response is too big, the server will return the first page that includes a part of the results. As the client intentions have been disclosed to the server, this allows for caching mechanisms to be set up on the server side.

Following the anonymous reviewer comments and the discussions during the InterOSS-IoT Workshop in Stuttgart on November 7th, 2016, authors will consider bringing this forward to OGC for consideration as a *white paper*.

To summarize, we argued that current SOS design was not intended for the Internet of the Things era. We designed a pagination extension offering progressive data transmission, economizing resources and tackling with limited or interrupted Internet connectivity with a disruption-tolerant protocol, while respecting SOS specification. There is a small effort into extending current SOS servers and clients to implement the pagination extension, while there are significant performance improvements, as indicated by the experimental results. The pagination extension sets the grounds for enabling SOS as an Internet of the Things dissemination outlet for sensor observations.

Supplementary Materials

Pagination cumulative results are available on Zenodo:
http://doi.org/10.5281/zenodo.178913

Acknowledgements. This work was partially supported by Greek General Secretariat for Research and Technology, grant 11SYN-6-411 (ALPINE), and the European Community's Seventh Framework Programme grant 613817 (MODEXTREME). Authors are grateful to the four anonymous reviewers for their valuable feedback and the participants of the InterOSS-IoT Workshop (Stuttgart, Nov. 7th, 2016) for their comments.

References

1. Alamdar, F., Kalantari, M., Rajabifard, A.: Towards multi-agency sensor information integration for disaster management. Comput. Environ. Urban Syst. **56**, 68–85 (2016). http://dx.doi.org/10.1016/j.compenvurbsys.2015.11.005
2. de Assis, L.F.F.G., Behnck, L.P., Doering, D., de Freitas, E.P., Pereira, C.E., Horita, F.E.A., Ueyama, J., de Albuquerque, J.P.: Dynamic sensor management: extending sensor web for near real-time mobile sensor integration in dynamic scenarios. In: Proceedings of International IEEE Advanced Information Networking and Applications (AINA), pp. 303–310, March 2016. http://dx.doi.org/10.1109/AINA.2016.100
3. Atzori, L., Iera, A., Morabito, G.: The Internet of Things: a survey. Comput. Netw. **54**(15), 2787–2805 (2010). http://dx.doi.org/10.1016/j.comnet.2010.05.010
4. Botta, A., de Donato, W., Persico, V., Pescapé, A.: Integration of cloud computing and Internet of Things: a survey. Future Gener. Comput. Syst. **56**, 684–700 (2016). http://dx.doi.org/10.1016/j.future.2015.09.021
5. Botts, M., Percivall, G., Reed, C., Davidson, J.: OGC® sensor web enablement: overview and high level architecture. In: Nittel, S., Labrinidis, A., Stefanidis, A. (eds.) GSN 2006. LNCS, vol. 4540, pp. 175–190. Springer, Heidelberg (2008). doi:10.1007/978-3-540-79996-2_10
6. Broering, A., Remke, A., Stasch, C., Autermann, C., Rieke, M., Möllers, J.: enviroCar: a citizen science platform for analyzing and mapping crowd-sourced car sensor data. Trans. GIS **19**(3), 362–376 (2015). http://dx.doi.org/10.1111/tgis.12155
7. Bröring, A., Echterhoff, J., Jirka, S., Simonis, I., Everding, T., Stasch, C., Liang, S., Lemmens, R.: New generation sensor web enablement. Sensors **11**(3), 2652–2699 (2011). http://dx.doi.org/10.3390/s110302652
8. Bröring, A., Janowicz, K., Stasch, C., Schade, S., Everding, T., Llaves, A.: Demonstration: a RESTful SOS proxy for linked sensor data. In: Proceedings of 4th International Workshop on Semantic Sensor Networks (SSN11), pp. 123–126 (2011)
9. Center for operational oceanographic products and services (co-ops) sensor observation service (2017). https://opendap.co-ops.nos.noaa.gov/ioos-dif-sos/. Accessed 22 Jan 2017
10. Curbera, F., Duftler, M., Khalaf, R., Nagy, W., Mukhi, N., Weerawarana, S.: Unraveling the web services web: an introduction to SOAP, WSDL, and UDDI. IEEE Internet Comput. **6**(2), 86 (2002). http://dx.doi.org/10.1109/4236.991449
11. Drosatos, G., Efraimidis, P., Athanasiadis, I., Stevens, M., D'Hondt, E.: Privacy-preserving computation of participatory noise maps in the cloud. J. Syst. Softw. **92**, 170–183 (2014). http://dx.doi.org/10.1016/j.jss.2014.01.035
12. Fredericks, J.J., Botts, M., Cook, T., Bosch, J.: Integrating standards in data QA/QC into OpenGeospatial consortium sensor observation services. In: Proceedings of OCEANS 2009-EUROPE, pp. 1–6, May 2009. http://dx.doi.org/10.1109/OCEANSE.2009.5278211
13. Goodchild, M.F.: Citizens as sensors: the world of volunteered geography. GeoJournal **69**(4), 211–221 (2007). http://dx.doi.org/10.1007/s10708-007-9111-y
14. Hashem, I.A.T., Yaqoob, I., Anuar, N.B., Mokhtar, S., Gani, A., Khan, S.U.: The rise of "big data" on cloud computing: review and open research issues. Inf. Syst. **47**, 98–115 (2015). http://dx.doi.org/10.1016/j.is.2014.07.006
15. Havlik, D., Bleier, T., Schimak, G.: Sharing sensor data with SensorSA and cascading sensor observation service. Sensors **9**(7), 5493–5502 (2009). https://dx.doi.org/10.3390/s90705493

16. Horita, F.E., de Albuquerque, J.P., Degrossi, L.C., Mendiondo, E.M., Ueyama, J.: Development of a spatial decision support system for flood risk management in Brazil that combines volunteered geographic information with wireless sensor networks. Comput. Geosci. **80**, 84–94 (2015). https://doi.org/10.1016/j.cageo.2015.04.001

17. Janowicz, K., Bröring, A., Stasch, C., Schade, S., Everding, T., Llaves, A.: A RESTful proxy and data model for linked sensor data. Int. J. Digit. Earth **6**(3), 233–254 (2013). http://dx.doi.org/10.1080/17538947.2011.614698

18. Jazayeri, M.A., Liang, S.H., Huang, C.Y.: Implementation and evaluation of four interoperable open standards for the Internet of Things. Sensors **15**(9), 24343–24373 (2015). http://dx.doi.org/10.3390/s150924343

19. Jirka, S., Bröring, A., Kjeld, P., Maidens, J., Wytzisk, A.: A lightweight approach for the Sensor Observation Service to share environmental data across Europe. Trans. GIS **16**(3), 293–312 (2012). http://dx.doi.org/10.1111/j.1467-9671.2012.01324.x

20. Lagoze, C., Van de Sompel, H.: The open archives initiative: building a low-barrier interoperability framework. In: Proceedings of 1st ACM/IEEE-CS Joint Conference on Digital libraries, JCDL 2001, pp. 54–62. ACM, New York (2001). http://doi.acm.org/10.1145/379437.379449

21. Li, S., Da Xu, L., Zhao, S.: The Internet of Things: a survey. Inf. Syst. Front. **17**(2), 243–259 (2015). http://dx.doi.org/10.1007/s10796-014-9492-7

22. Mulligan, G., Gracanin, D.: A comparison of SOAP and REST implementations of a service based interaction independence middleware framework. In: Proceedings of Winter Simulation Conference (WSC), pp. 1423–1432, December 2009. http://dx.doi.org/10.1109/WSC.2009.5429290

23. Observations and Measurements - XML implementation. Implementation Standard 10–025r1, Open Geospatial Consortium (2011)

24. OGC Sensor Observation Service 2.0. Implementation Standard 12–006, Open Geospatial Consortium (2012)

25. OGC Sensor Observation Service 1.0. Standard 06–009r6, Open Geospatial Consortium (2007)

26. OGC Sensor Observation Service 2.0 Hydrology Profile. Best Practice Paper 14–004r1, Open Geospatial Consortium (2014)

27. SensorML, O.G.C.: Model and XML. Encoding Standard 12–000, Open Geospatial Consortium (2014)

28. OGC SensorThings API Part 1: Sensing. Implementation Standard 15–078r6, Open Geospatial Consortium (2016)

29. OGC Web Feature Service 2.0. Interface Standard 09–025r2, Open Geospatial Consortium (2014)

30. Papazoglou, M., Georgakopoulos, D.: Service-oriented computing. Commun. ACM **46**(10), 25 (2003). https://doi.org/10.1145/944217.944233

31. Perera, C., Zaslavsky, A., Christen, P., Georgakopoulos, D.: Sensing as a service model for smart cities supported by Internet of Things. Trans. Emerg. Telecommun. Technol. **25**(1), 81–93 (2014). https://doi.org/10.1002/ett.2704

32. Pradilla, J., Palau, C., Esteve, M.: SOSLITE: lightweight Sensor Observation Service (SOS) for the Internet of Things (IoT). In: ITU Kaleidoscope: Trust in the Information Society (K-2015), pp. 1–7. IEEE, December 2015. https://doi.org/10.1109/Kaleidoscope.2015.7383625

33. Regueiro, M.A., Viqueira, J.R., Taboada, J.A., Cotos, J.M.: Virtual integration of sensor observation data. Comput. Geosci. **81**, 12–19 (2015). http://dx.doi.org/10.1016/j.cageo.2015.04.006

34. Reitz, K.: Requests: HTTP for humans (2017). http://docs.python-requests.org/en/master/. Accessed 22 Jan 2017
35. Rouached, M., Baccar, S., Abid, M.: RESTful sensor web enablement services for wireless sensor networks. In: IEEE Eighth World Congress on Services, pp. 65–72. IEEE, June 2012. https://doi.org/10.1109/SERVICES.2012.48
36. Samourkasidis, A., Athanasiadis, I.N.: A miniature data repository on a Raspberry Pi. Electronics **6**(1) (2017). http://dx.doi.org/10.3390/electronics6010001
37. The Python standard library: random - Generate pseudo-random numbers. (2017). https://docs.python.org/2/library/random.html. Accessed 22 Jan 2017
38. The Python standard library: time - time access and conversions (2017). Accessed 22 Jan 2017. https://docs.python.org/2/library/sqlite3.html
39. Yazar, D., Dunkels, A.: Efficient application integration in IP-based Sensor networks. In: Proceedings of 1st ACM Workshop on Embedded Sensing Systems for Energy-Efficiency in Buildings, BuildSys 2009, pp. 43–48. ACM, New York (2009). http://doi.acm.org/10.1145/1810279.1810289

Requirement-Based Deployment of Applications in Calvin

Ola Angelsmark[(✉)] and Per Persson

Ericsson Research Cloud Technology, Mobilvägen 1, 223 62 Lund, Sweden
{ola.angelsmark,per.persson}@ericsson.com

Abstract. In order for IoT application developers to deliver on the promise of IoT, new tools and methodologies addressing the challenges associated with development of highly distributed systems running on non-reliable and heterogeneous hardware are required. Some of the main characteristics of cloud computing that has been a driving force for its success, are resource pooling, elasticity and the capacity for combining unrelated services. We believe that a similar approach is needed for IoT as well. In this paper, we show how Calvin, an open source peer-to-peer platform for distributed applications, tackles many of the problems inherent in IoT. By only loosely associating the functionality of a device with a semantics, and having implementations based on conventions rather than pre-defined terminology, it is possible to let a system of Calvin runtimes autonomously handle deployment decisions, and respond to changing requirements. We will discuss how to develop and deploy dynamic and adaptive IoT-applications based on capabilities and requirements, and how to resolve requirements by automatically combining information from multiple sources based on encapsulated domain knowledge.

1 Introduction

There is no doubt that IoT solutions are required within our society to increase efficiency and effectiveness in industry and to help cities and communities achieve the necessary economic growth with reduced impact on the environment. Without effective means for programmers to develop applications, however, the much needed IoT revolution will progress slower and have less impact than is desired.

Some of the main characteristics of cloud computing that has been a driving force for its success, are resource pooling, elasticity and metering [20]. Physical resources such as compute and storage nodes, and network fabrics are shared among tenants. Virtual resource elasticity brings the ability to dynamically change the amount of allocated resources, for example as a function of workload or cost. Resource usage is metered and in most pricing models the tenant only pays for allocated capacity. We believe that a similar approach is needed for IoT as well.

Virtualization techniques have played a major role in the success of cloud computing through simplifying resource sharing and providing isolation. Similarly, for the IoT scenario, we could benefit from sharing sensors, actuators, and

I. Podnar Žarko et al. (Eds.): InterOSS-IoT 2016, LNCS 10218, pp. 72–87, 2017.
DOI: 10.1007/978-3-319-56877-5_5

compute nodes between applications and application domains. Traditional virtualization techniques, such as virtual machines and containers, are associated with a certain amount of overhead and are slow to migrate. A virtualization platform for IoT must be extremely lean and able to execute on devices with very scarce resources.

Like cloud resources, things in IoT, such as sensors and actuators should not be hardwired to any particular application, but instead be viewed as parts of a generic computing platform. This decouples application development and deployment from hardware investments. For example, sensors in vehicles are of interest to many parties, and could be shared among users. An obvious application would be to open up the outside thermometer of cars as a way of crowd sourcing weather data, but there are numerous other possibilities, e.g., deducing traffic jams based on speed and location of vehicles on a road. This naturally raises a number of interesting issues on its own, such as how this can be done without compromising security and privacy. These topics are broad enough to fill several papers on their own, and are mostly outside of the scope of the current paper. See, e.g. [12, 22] for a discussion on the security aspects in a distributed IoT framework such as Calvin. Privacy preserving secure resource sharing, on the other hand, is still a topic for future research.

In order for IoT application developers to deliver on the promise of IoT [5], new tools and methodologies that address the challenges associated with the development of highly distributed systems running on a non-reliable and heterogeneous hardware platform are required.

To address this, we propose to combine, and refine, the well-known actor and data-flow programming models to achieve true separation of concerns between (at least) four entities: (1) device manufacturers, (2) developers, (3) market places that hosts applications and components, and (4) operators that deploys and maintains applications.

The rest of this paper is laid out as follows: First we paint a picture of our vision, and then we will look at some successful uses of the strategies we combine, and lay out our strategy in some detail before we turn to presenting our solution platform. Then we will look at a scenario where the properties of the platform is used to create IoT applications with novel properties, and finally we summarize the results and talk about where we are heading next.

2   A Vision of Software Definable Things

In a scenario when everything is programmable, from the smallest sensors to the servers in cloud and the nodes in the communication networks, we need a new way to model and implement IoT applications. We envision a developer methodology that provides a unified programming view of such a heterogeneous execution platform. From the programmer's point of view, all computing resources should be treated in an abstract manner, independent of whether it will eventually reside on a server in a data centre or on an embedded sensor. An important piece to the puzzle is the management of resources. A holistic approach, which maximizes the utilization of hardware at lower cost while at the same time maintaining timing and other types of requirements, as well as properties particular to each specific device, is required. We are not alone in arriving at these conclusions, see e.g. [11].

From this point of view, current state-of-the-art in IoT development is, mostly, (1) cloud centric with dumb devices for sensing and actuating, with application logic residing in the cloud, (2) siloed, preventing reuse of existing infrastructure, (3) insufficiently abstracted, requiring developers to specify IP-addresses, transport protocols, and device parameters, (4) assuming deployments to be static, and (5) too programming oriented, requiring a complex code just to get started with simple applications. Much of this is caused by viewing IoT-applications as just another kind of cloud application, and while we should keep the properties of cloud computing that are beneficial, we clearly need to adapt any good solution to the peculiarities of IoT.

We need to describe applications in an abstract manner, regardless of the details of hardware, and the actor model [13] provides a clear separation of concerns by encapsulating functionality and state, and providing well defined interfaces, *ports*, towards other actors. Furthermore, from a software engineering perspective, actors promote reuse, opening up the possibility for making them available through an open repository or marketplace.

In an application the actors have to be connected such that data flows from port to port, and in the CAL actor language [7], an application is described as a directed graph of interconnected actors, a data-flow. This description captures the functional behaviour of the application, but does not prescribe exactly *how* the application should be deployed. Typically, an application may be deployed or orchestrated in multiple ways, which opens up for different kinds of optimization.

Central to achieving dynamic application deployment and management is matching an application's requirement against host capabilities and other attributes. For example, in a building automation scenario, a sensing actor may state a requirement to be deployed on any device capable of measuring temperature within a specified building and within a particular security domain. The programming model must thus provide means for elaborate specification of requirements and capabilities.

As the field of IoT is rapidly evolving we believe that a successful platform must be an open one, and there are several ways of creating an open platform. One way would be to build a world-wide standard for all types of sensors, actuators, and other kinds of IoT devices. This would require extensive taxonomies and a more or less fixed enumeration of capabilities and classifications. We believe a more agile process is preferable, implementing support for devices as and when it is needed — learning what works and what does not while doing so — being open to, and try to recognize, better solutions when they appear. In keeping with the Agile Manifesto [2], we prefer a good solution to the problem at hand over a general solution to a problem we have not yet encountered. Once a useful de facto standard has crystallized, however, there is of course no reason not to make an official standard out of it.

In our proposed scenario, we see all hosts, from the smallest IoT devices, to base-stations and cloud resource, hosting, or being proxied by, a runtime execution environment. The runtime takes care of establishing communication and sending messages between runtimes over available network interfaces. It also offers a Hardware Abstraction Layer (HAL) for applications to interact with e.g. sensors or actuators, to establish a common interface to similar functionality, built up in an agile manner, meaning we see it as evolving over time, with developers of similar functionality eventually agreeing on a de facto standard. This is discussed further in Sect. 4.2.

Note that by 'runtime' we mean software handling the hardware abstraction, communication with peer runtimes, local resource management, etc. It *could* be implemented as a virtual machine executing bytecode, but this is not a requirement. As long as the runtime can be part of the system, i.e. it implements the protocol of the reference implementation, then it does not really matter how it is implemented. If a device has a structured way of describing and handling resources, such as IPSO [17], then it is most likely better to build a runtime on top of this, rather than replace it.

Applications, described as directed graphs with self-contained actors as nodes, are deployed to such runtimes. The application approach is similar to a micro-service architecture common in modern cloud applications, with self-contained services connected to make up the application. An application typically spans over many hosts. Similarly, each host executes actors from multiple applications, possibly from different tenants.

The Calvin platform [23] is an attempt at a solution allowing developers to develop applications using clearly separated, well-defined functional units (actors) and per-deployment requirements. The platform then autonomously manages the application by placing the actors on different nodes (devices, network nodes, cloud, etc.) in order to meet the requirements, and later migrates them if changes in circumstances should so require.

3 Related Approaches

In this section we briefly present a non-exhaustive list of other frameworks that are to some extent based on, or at least influenced by, the same ideas as Calvin.

NoFlo [3] and Node-RED [15] are two frameworks exploiting the simplified development possible with the Flow-based programming model [21]. By specifying how data moves through an application, together with well-defined interfaces between components, it is possible to make application development quite intuitive and simple, for example by using a graphical drag-and-drop interface.

The Actor model of computation is exploited in the Orleans framework [6] in order to get a lightweight virtualization which opens up a set of new possibilities. By keeping the internal state of an actor private, and only allowing it to change in response to events and data, it is possible to handle tasks such as scaling, migration and load balancing on a fine-grained level not possible using standard virtualization methods.

Lee *et al.* [19] discusses the issues and requirements needed to handle application development and deployment in large scale, heterogeneous, and dynamic systems. Applications in this setting, called swarmlets, uses actors to separate devices and services from the functionality they offer. By including a list of requirements in the description of an actor, it is possible to determine the functionality required for deploying an application beforehand.

Asynchronous, event-driven architectures are well suited for loosely coupled components and services [9]. On the highest level, it is also easy to visualize the causality in the application, which, as is the case with the dataflow model, makes application development manageable. The approach has been very successful in cloud-based frameworks, such as IFTTT [16], and OpenWhisk [14], as well as the AWS Lambda [1] framework. To some extent, they all try to make the binding between triggering events and consequent actions as easy and transparent as possible. For IFTTT, it is done by combining pre-defined, more or less well-established services, whereas for OpenWhisk and AWS Lambda, the focus is more on how to develop and combine services.

The natural question is of course what sets Calvin apart from the aforementioned platforms, or indeed any of the many other very competent frameworks that exist today. Calvin differs from other platforms in three main ways: (1) It is peer-to-peer, allowing for greater flexibility in where computations are made and decisions taken, (2) the separation of concerns between the different stages of the application lifecycle lets each stakeholder focus on their own specialties, and finally (3) the requirement-based deployment of applications allows for a greater level of automatic management, which is the topic of this paper.

4 Calvin

The development of the open source Calvin platform [8] was initiated and is actively maintained by Ericsson Research, based on previous experience working with dataflow and actor-based paradigms for multicore systems [4]. The basic idea behind this platform is that a distributed system is very similar to a multicore system, but with heterogeneous components, and large differences in throughput and latency between them.

One of the goals the of Calvin is to simplify developing distributed applications. In order to facilitate this, the individual runtimes combine to present a unified front to the application, and the application developer, giving the appearance of a single runtime offering the resources of the entire system, and, as a consequence, applications can be written as were they run-of-the-mill, non-distributed apps, which greatly simplifies their development. This poses some limitations on the programming paradigm; this approach is not suitable for e.g. imperative programming methodologies. In the following sections, we will discuss this more in-depth.

The platform decouples application development and deployment from hardware investments by providing an abstraction layer for applications and establish a common interface to similar functionality, built up in an agile manner.

Part of what makes the Calvin platform so appealing is that it handles all communication in an application. This means that the developer does not have to know how a specific device communicates, or which protocols it uses. Provided the Calvin runtime has a means of communicating with it, it can be part of the system, preferably by hosting a Calvin runtime, or for legacy devices by having Calvin handling all communication, acting as a *proxy*, see Sect. 4.2, giving the appearance of the device being part of the system.

4.1   Calvin Application Development

We will now describe in more detail how Calvin makes it possible to handle large portions of deployment and management automatically. Much of the power and flexibility stems from dividing an application's life cycle into four separate, well defined, phases:

- *Describe.* In the first phase, the fundamental building blocks, actors, describing device functionality and common operations are created. In practice, there are actors that may execute on any runtime, actors providing services that requires resources not available everywhere, and sensing and actuating actors that require specific hardware to function. Actors are collected in a system wide actor store, available for reuse, and it is typically in the interest of device suppliers to provide device-specific actors.
- *Connect.* In the second phase, an application developer specifies actors and how data should flow between them. This gives an abstract view of the application, in effect capturing the essence of what the application does, but it does not say anything about which devices should be involved.
- *Deploy.* In the third phase, an application is deployed by mapping actors onto things, where a "thing" can be anything Calvin can communicate with. The mapping causes an application to be distributed across a set of devices and services matching its requirements.
- *Manage.* Finally, once deployed, the application enters its managed phase. Here, the application can be monitored and traced. Actors can be moved from one device to another in order to balance or optimize resource usage, and actors may even replaced while the application is running.

4.2   The Calvin Actor

As previously stated, Calvin uses the actor model to provide abstractions of device and service functionality. The actors are event-driven, and can be used to build loosely coupled applications, with each actor providing some functionality the application needs. Figure 1 gives an overview of a Calvin actor. Each actor is self-contained and hides its internal state from the rest of the application.

In Calvin, the standard model has been extended with a collection of metadata associated with each actor. This metadata contains information on the actor that the platform make use of during deployment and management. The information includes, for example, requirements the actor have on the runtime.

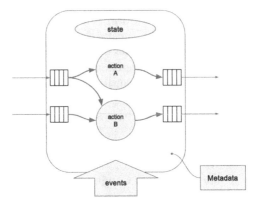

Fig. 1. A Calvin actor.

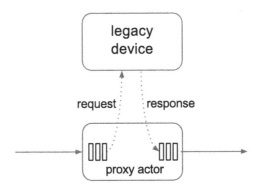

Fig. 2. A Calvin proxy actor.

In order to migrate an actor, the runtime currently hosting it serializes the internal state and sends it to the destination runtime, which then creates a new instance of the actor, restoring state from the serialized data. It is important to note that implementation details of the actor (and runtime) are irrelevant for the migration. As long as the received state can be used, either directly or after some transformation, then the actor can be migrated to this runtime. Naturally, when transforming the state, great care must be taken so that it is not mangled in a way which prevents it from being migrated to another runtime.

Since a migration entails creating a new instance of an actor initialized with the old state, this mechanism allows for easy upgrade of an actor in a running system. By migrating an actor to the same runtime, and using a new version of the desired actor, initialized with the old state, an upgrade can be done with minimal impact. Migration can also be used when upgrading the Calvin runtime itself — start a new, upgraded runtime, migrate all actors to it, and then stop the old one. Should there be insufficient resources to handle two runtimes on a single device, then it is also possible to migrate all actors to an intermediary runtime elsewhere during the upgrade.

In order to handle legacy devices and services, i.e. those that for some reason cannot host a runtime, a common pattern is to use a special kind of actor, a proxy actor. As shown in Fig. 2, a proxy actor translates an incoming message on a port to a request to an external legacy device, using whatever API or protocol exposed by the device, and subsequently turns the response into an outgoing message from the actor. This is similar to the *Accessor* pattern described by Latronico *et al.* [18].

4.3 CalvinScript

With actors describing the processing blocks, we can express the data flow graph in a concise manner using *CalvinScript*, a small, domain specific description language. An example of an application written in CalvinScript is shown in Listing 1.1.

Listing 1.1. A simple application

```
trigger : std.Trigger(tick=1, data=null)
camera  : media.Camera()
screen  : media.ImageRenderer()
status  : io.Print()

trigger.token > camera.trigger
camera.image > screen.image
camera.status > status.token
```

The application comprises four actor instances, `trigger`, `camera`, `screen`, and `status`, defined at the top, followed by a description of the data flow. The flow of data is stated as *actor.outport > actor.inport* where the > operator denotes data flow from outport to inport, not an ordering in the mathematical sense.

4.4 Capabilities and Requirements

In order to handle the complexity of systems with a large number of devices, deployment and management should be orchestrated by the runtimes themselves, without human interaction. The way this is handled in Calvin, is through matching a collection of *requirements* an application has with the *capabilities* presented by the system, where capabilities (somewhat simplified) represent what a runtime (and thus a system) offers in what it can do, and its properties. In other words, a Calvin runtime presents an abstraction of the platform it runs on as a collection of capabilities this platform offers. Analogously, requirements represent what the application needs in order to function.

Capabilities. Capabilities are fetched from the device or platform the runtime is executing on, but they can also be provided as a configuration to the runtime.

Besides the fundamental capabilities of a device, e.g. that a camera provides images, we also consider information related to location, ownership etc. as a kind of capability. An example is shown in Listing 1.2 where a physical camera is tied to its particular location by adding a piece of information to the runtime, typically when installing the camera.

The capabilities are shared between runtimes in a registry, and while the internal API to access the registry is prescribed, the implementation is not. Currently, Calvin comes with two different implementations for storing capabilities; a central repository, and a secure distributed hash table (DHT), and it is a matter of configuration to choose one of them.

Listing 1.2. Additional capabilities for a particular device (partial)

```
{
  "indexed_public": {
    "address": {
      "country":"SE",
      "locality":"Lund",
      "street":"Testvagen",
      "streetNumber":"1",
      "room":"secret_room"
    },
    "node_name": {
      "organization":"com.ericsson",
      "name":"secret_room"
    }
  }
}
```

Requirements. Almost all actors have requirements on the hosting runtime, such as the presence of a timer, the ability to measure temperature, or access to a file system. Once a developer has selected a specific actor when writing an application, the requirements of this actor become implicit to the application, and cannot be changed (unless the actor is replaced by one with different requirements.) Consequently, the selection of actors can have a major impact on which runtimes can host the application. At first glance, this may seem as a limitation, but it means that it is straightforward to collect the requirements for an application, and determine whether the application can run on a given system or, if not, give an explanation of what is missing. The developer can then either change the application or add the missing capabilities to the system.

In addition to the requirements stemming from actors, there are additional requirements posed by the applications, usually given by the developer (or whoever is in charge of deploying and maintaining the application.) These requirements include properties such as geographical location, ownership, or name of a runtime. With this, it is possible to specify a *particular* runtime (device) to host an actor, or that the runtime (device) should reside in a certain location. As an example, the deployment requirements for the application in Listing 1.1 could, in one particular scenario, be as shown in Listing 1.3. Note that we do not specify

Listing 1.3. Deployment requirements (partial)

```
{
  "requirements": {
    "camera": [{
      "op":"node_attr_match",
      "kwargs": {"index":
        ["address", {"country":"SE","locality":"Lund",
                     "street":"Testvagen",
                     "streetNumber":"1"}]
      },
      "type":"+"
    }],
    "screen": [{
      "op":"node_attr_match",
      "kwargs": {"index": ["node_name", {"name":"laptop"}]},
      "type":"+"
    }],
  }
}
```

any particular room, but rather just the general address of the premises. Thus, any room or location at that address will match the deployment requirements.

Security as Capabilities and Requirements. Although we are not addressing Calvin authentication and security in this paper, it is worth noting that access management and access control is also treated as collections of requirements and capabilities; the application is deployed and executes as a user in the system, and in order for an actor to execute on a runtime, this user needs first of all to be allowed to place an actor on this runtime, and, second, have access to the capabilities satisfying the requirements of the actor. The details of this are discussed further in [22].

Given the surveillance application from Listing 1.1 and the deployment requirements from Listing 1.3 we now have a surveillance system that accepts input from each possible camera and displays the stream on a computer screen.

4.5   Deployment

As stated in Sect. 4, Calvin simplifies application development by giving the illusion of a single runtime, but of course the application *is* distributed and pushing the complexity downwards does not solve the problems associated with distributed applications per se. It does however confine them to the core layers of Calvin, where they can be handled systematically. The solution relies heavily on the capabilities and requirements described in Sect. 4.4, the hardware abstraction layer, called *CalvinSys*, and the abstraction and hiding of transport and communication protocols. All of these work in concert during deployment, and when adapting to changing conditions during an application's life, to migrate actors to appropriate runtimes while ensuring data flow between actors.

Deployment of a Calvin application basically works as follows: The application graph of actors and connections, together with any additional requirements, are handed over to a runtime in the system. The runtime goes through the list of actors, determine their type, possibly by looking it up in the registry, instantiates them and sets up the port connections. If an actor has unsatisfied requirements, it will not be able to execute on this runtime, and is kept in a *shadow* state for now. Next the runtime looks at the requirements, both the implicit actor requirements and the supplied application requirements, querying the registry for runtimes in the system with matching capabilities. Once it has found a matching runtime for each actor in the application, the actors are migrated to their initial deployment and the application can start executing.

Of course, if all requirements are satisfied by the initial runtime, then there will be no migration, and the application can commence executing at once. Should there, on the other hand, be requirements that cannot be fulfilled, there will be actors left in a shadow state, and the application cannot start.

This approach to deployment and functionality offers the possibility of having a runtime handle an actor and its requirements in a "best effort" fashion. This is done by having the runtime present level of abstraction centered around the functionality, rather than the hardware. So, for example, a runtime can expose an "alarm" functionality, without specifying exactly what the alarm does, and an application can make use of this without the developer having detailed knowledge it. When an alarm is triggered, the actor gets a token on a port containing relevant data in order of priority — such as "audio & video," "just audio," "text," or simply "on". The runtime hosting the actor then executes its alarm functionality, using the available data. For example, an alarm on a system with a tv monitor could use audio and video to display and play a message, a speaker could use the audio stream, and a small, cellular device could send the text as a text message to a preconfigured recipient. There could also be a predefined sequence of actions being executed by the device in response to an alarm, disregarding any data in the call.

4.6 Management

One of the nice properties of the automated deployment based on capabilities and requirements is that many aspects of managing running applications are already in place. By making sure that changes in conditions and capabilities triggers the same logic that handles initial deployment, the correct measures to ensure that a running application's requirements are met are taken. Typically these measures result in one or more actors migrating from one runtime to another.

5   Scenario: No Access at Certain Times

Next we present a scenario where we utilize requirements and capabilities to achieve goals that are not easily handled by current IoT-systems.

Coming back to our surveillance scenario from Sect. 4.5, we would like to prevent cameras from accessing the top secret lab during office hours, while allowing security and management staff to be able to view those streams during off-hours to make sure no-one is in the lab. We can do this by amending the general policy with the policy given in Listing 1.4 specifically for the lab.

The nice thing about policy, and security in general, is that to Calvin it behaves as any other requirement. Thus, by applying the policy in Listing 1.4 we prevent migration of the camera actor to the secret lab during office hours using the same mechanisms used for satisfying functional requirements, and in case the camera actor is currently executing on the camera in the secret room, it is migrated away from that runtime. An in-depth description of how authorization is handled in Calvin is given by Nilsson [22].

Listing 1.4. Policy for secret room

```
{
  "id":"policy1",
  "rule_combining":"permit_overrides",
  "target": {"resource": {"node_name.name":"secret_room"}},
  "rules": [{
    "id":"policy1_rule1",
    "effect":"permit",
    "condition": {
      "function":"or",
      "attributes": [
          {"function":"equal",
          "attributes": ["attr:subject:group","Security"]},
          {"function":"equal",
          "attributes": ["attr:subject:position","Manager"]}
      ]
    },
    "obligations": [{
      "id":"time_range",
      "attributes":{"start_time":"16:00","end_time":"08:00"}
    }]
  }]
}
```

6   Next Steps

The functionality necessary for the applications and deployments discussed so far are already in place in Calvin, and although it is sometimes cumbersome to express e.g. requirements, it is possible to run the examples. However, there are a number of future extensions that offer some quite interesting possibilities.

Listing 1.5. Actor encapsulating domain knowledge (informal pseudo-code)

```
@provides(roadspeed)

@requires(powertrain.roadspeed)
return read(powertrain.roadspeed)

@requires(powertrain.distance, dashboard.real_time_clock)
define comp {
    t0, s0 := 0, 0 // initial values
    t = read(dashboard.real_time_clock)
    s = read(powertrain.distance)
    v = (s-s0)/(t-t0)
    s0, t0 = s, t
}
return comp()
```

Actor Enhancements. An interesting possibility is to include insights and learnings such as historically successful (or unsuccessful) deployments in the state of an actor. For example, if, for some reason, it is the case that a particular actor always performs poorly on a specific runtime, even though there are no obvious reasons for this, that information can be used during future deployments in order to avoid this runtime.

Encapsulated Domain Knowledge. Consider the case when a new capability has been introduced, and has become popular, but it is not yet widely supported among existing runtimes. The capability can in that case be implemented in an actor, which includes a prioritized list of capabilities it could potentially make use of in order to run. (The ability to have conditional inclusions on requirements is an upcoming, but not currently supported feature in Calvin.)

This is in some sense encapsulation of domain knowledge within an actor, and can be used to, e.g., have a "backup" plan for an actor. A trivial, but illustrative, example is shown in Listing 1.5. The actor calculates the roadspeed of a vehicle. The preferred way is to read the current speed from the powertrain, but if this sensor is not available, then the actor can make use of two different inputs in order to calculate the speed, using the well-known formula $\text{speed} = \frac{\text{distance}}{\text{time}}$.

Soft Requirements, Incremental Deployments. A requirement can be either *hard*, meaning it is non-negotiable and must be met in order for the application to execute at all, or *soft*, which means the application can function without it, but it will perform better, in some way, if it is satisfied. Calvin currently only supports hard requirements — the plan is to add support for soft requirements in some form within the year.

Also in development are application requirements involving several actors. For example, in an application involving streaming data, there can be requirements on

Listing 1.6. A small video application

```
src:  media.Camera()
snk:  media.ImageRenderer()


true > src.trigger // Continuous trigger
src.image > snk.image
snk.status > voidport // Discard status output
```

bandwidth or latency in between two or more actors. Such a requirement is best satisfied by having all actors involved hosted by the same runtime, but if this is not possible, due to e.g. resource constraints or lack of computational power, then the requirement extends to all runtimes hosting these actors. In general, this is a problem related to the Constraint Satisfaction Problem which is known to be intractable [10], but provided the requirements are soft, thus allowing the application to execute without necessarily having all requirements met, there are ways of handling this using Calvin.

As an example, consider the application in Listing 1.6, slightly modified from Listing 1.1. This application, as is, sends images as fast as possible from a camera to a screen. A typical soft requirement to add here is the framerate of the stream. With Calvin, the way to handle this will be to do a first deployment which does not fully satisfy the framerate requirement. It could even be a deployment where it is unknown what the framerate will be. As the runtimes "learn" their frame rate capabilities, they can make decisions on where to migrate the actors in the application in order to, incrementally, improve the framerate of the stream. For an application in a large system, this can take a significant number of migrations, but the application will always be running, albeit at a reduced framerate.

Broadcast and Covering Migrations. For many applications, the goal is simply to collect data from a multitude of devices and send it to be analyzed in real-time, or stored for later analysis. These applications can be greatly simplified using a planned feature which allows an actor to be replicated on a number of runtimes based on e.g. an attribute or a capability. For example, an application for collecting the temperature from all thermometers in a building could be (somewhat simplified) described as in Listing 1.7, with the added requirement that the actor thermometer (with the obvious interpretation) should be present on one runtime in each room in the building. It is not important *which* specific thermometer (or other device) is used in each room as long as one is. The actor measurement, which is of type attr.TagData, an actor which is used to tag data with a given attribute, in this case address.room, would have as an additional requirement that it is hosted by the same device as the thermometer.

An extension to the functionality from the previous paragraph is the addition of a "broadcast" mode of migration, where an actor is sent to *all* runtimes matching some requirement. For example, an emergency vehicle travelling at high speed towards the scene of an accident could have an emergency application which is

Listing 1.7. A fragment of a temperature application

```
// Actors
fire: std.Trigger(tick=0.5, data=true)
measurement: attr.TagData(index="address.room")
thermometer: sensor.Temperature()

// Connections
fire.data > thermometer.measure
thermometer.centigrade > measurement.data
measurement.data > ...
```

broadcast with the requirements that the receiving runtimes be located along a certain path, and be either a vehicle or a traffic light controlling an intersecting road. By having the deployment requirement be, for example, all devices in the building, it would suffice to deploy a single application and broadcast migrate. Less critical applications that could make use of this would be e.g. applications for subscribing to sports results or news.

7   Summary and Conclusions

With Calvin, we hope to address some of the many issues that are holding back IoT, preventing it from deliver on its promises. Moving away from a cloud centric model with siloed deployments, we want to move the intelligence from the data center to the devices, and let them talk directly to each other, making decisions in concert, thus necessarily breaking open the silos.

We think it is clear that we are on the right track. The ease with which even quite hard problems, such as dynamic deployment and automatic management of applications can be modelled and implemented using requirements and capabilities, shows that the ideas are viable, both in theory and in practice.

References

1. Amazon Web Services Inc.: AWS Lambda — Serverless Compute (2016). https://aws.amazon.com/lambda/
2. Beck, K., Beedle, M., van Bennekum, A., Cockburn, A., Cunningham, W., Fowler, M., Grenning, J., Highsmith, J., Hunt, A., Jeffries, R., Kern, J., Marick, B., Martin, R.C., Mellor, S., Schwaber, K., Sutherland, J., Thomas, D.: Manifesto for Agile Software Development (2001). http://www.agilemanifesto.org/
3. Bergius, H.: NoFlo — Flow-based Programming for Javascript (2015). http://noflojs.org
4. Bhattacharyya, S.S., Brebner, G., Janneck, J., Eker, J., von Platen, C., Mattavello, M., Raulet, M.: Opendf: a dataflow toolset for reconfigurable hardware and multi-core systems. SIGARCH Comput. Archit. News. **36**(5), 29–35 (2009)
5. Biddlecombe, E.: UN predicts "Internet of Things" (2005). http://news.bbc.co.uk/2/hi/technology/4440334.stm

6. Bykov, S., Geller, A., Kliot, G., Larus, J., Pandya, R., Thelin, J.: Orleans: a framework for cloud computing. Technical report MSR-TR-2010-159, Microsoft Research (2010)
7. Eker, J., Janneck, J.: CAL language report: specification of the CAL actor language. Technical memorandum UCB/ERL M03/48, University of California Berkely (2003)
8. Ericsson Research: Calvin — Lets Things Talk to Things (2015). https://www.github.com/EricssonResearch/calvin-base
9. Eugster, P.T., Felber, P.A., Guerraoui, R., Kermarrec, A.M.: The many faces of publish/subscribe. ACM Comput. Surv. (CSUR) **35**(2), 114–131 (2003)
10. Garey, M.R., Johnson, D.S.: Computers and Intractability: A Guide to the Theory of NP-completeness. W.H. Freeman, New York (1979)
11. Giang, N.K., Blackstock, M., Lea, R., Leung, V.C.M.: Developing IoT applications in the fog: a distributed dataflow approach. In: 2015 5th International Conference on the Internet of Things (IOT), pp. 155–162 (2015)
12. Gil, J.M.R.: Secure Domain Transition of Calvin Actors. Master's thesis, Lund University (2016)
13. Hewitt, C., Bishop, P., Steiger, R.: A universal modular actor formalism for artificial intelligence. In: Proceedings of the 3rd International Joint Conference on Artificial Intelligence. pp. 235–245. Morgan Kaufmann Publishers Inc. (1973)
14. IBM: Cloud-first distributed event-based programming service (2015). https://developer.ibm.com/openwhisk/
15. IBM Emerging Technologies: A visual tool for wiring the Internet-of-Things (2016). http://nodered.org/
16. IFTTT Inc: IFTTT — Put the internet to work for you (2015). https://ifttt.com
17. Jimenez, J., Koster, M., Tschofenig, H.: IPSO smart objects. In: IoT Semantic Interoperability Workshop 2016 (IOTSI) (2016)
18. Latronico, E., Lee, E.A., Lohstroh, M., Shaver, C., Wasicek, A., Weber, M.: A vision of swarmlets. IEEE Internet Comput. **19**(2), 20–28 (2015)
19. Lee, E.A., Hartmann, B., Kubiatowicz, J., Rosing, T.S., Wawrzynek, J., Wessel, D., Rabaey, J.M., Pister, K., Sangiovanni-Vincentelli, A.L., Seshia, S.A., Blaauw, D., Dutta, P., Fu, K., Guestrin, C., Taskar, B., Jafari, R., Jones, D.L., Kumar, V., Mangharam, R., Pappas, G.J., Murray, R.M., Rowe, A.: The swarm at the edge of the cloud. IEEE Des. Test **31**(3), 8–20 (2014)
20. Mell, P., Grance, T.: The NIST definition of cloud computing. Technical report, National Institute of Standards and Technology (2011)
21. Morrison, J.P.: Flow-Based Programming: A New Approach to Application Development. CreateSpace, Paramount (2010)
22. Nilsson, T.: Authorization Aspects of the Distributed Dataflow Oriented IoT Framework Calvin. Master's thesis, Lund University (2016)
23. Persson, P., Angelsmark, O.: Calvin — Merging Cloud and IoT. Procedia Comput. Sci. **52**, 210–217 (2015)

Business Models and Security

Business Models for Interoperable IoT Ecosystems

Werner Schladofsky[1(✉)], Jelena Mitic[2], Alfred Paul Megner[1], Claudia Simonato[3], Luca Gioppo[3], Dimitris Leonardos[4], and Arne Bröring[2]

[1] Atos IT-Solutions and Services GmbH, Vienna, Austria
{werner.schladofsky,alfred-paul.megner}@atos.net
[2] Siemens AG, Munich, Germany
{jelena.mitic,arne.broering}@siemens.com
[3] CSI-Piemonte, Turin, Italy
{claudia.simonato,luca.gioppo}@csi.it
[4] Econais, Patras, Greece
dleonardos@wubby.io

Abstract. The Internet of Things (IoT) is growing and more and more devices, so-called "things", are being connected every day. IoT platforms provide access to those "things" and make them available for services and applications. Today, a broad range of such IoT platforms exist with differing functional foci, target domains, and interfaces. However, to fully exploit the economic impact of the IoT, it is essential to enable applications to interoperate with the various IoT platforms. The BIG IoT project aims at enabling this interoperability and supporting the creation of vibrant IoT ecosystems, which facilitate the development of cross-platform and cross-domain applications. While the value of interoperability for the overall economy is well understood and cannot be underestimated, some stakeholders may still need to find their business value in interoperable IoT ecosystems. Thus, this paper identifies the different stakeholders of such ecosystems, and analyses how these stakeholders can enhance their existing business models when taking part in an interoperable IoT ecosystem.

Keywords: Internet of Things · Ecosystems · Business models · Interoperability

1 Introduction

Since its very beginnings, the notion of the "Internet of Things" (IoT) [1], as technology that enables physical assets to become parts of information chains, has experienced an ever increasing attention. Today, the IoT has become a reality for businesses and consumers. Connected devices, or "things", are the fundament of the IoT, and they range from connected light bulbs, over personal fitness trackers, to geolocated shipping containers. Various studies predict significant growth of the IoT and its business value in the coming years. E.g., Gartner anticipates an increase from 6 billion connected devices in 2016 to over 20 billion in 2020 [2]. A recent McKinsey analysis [3] foresees that, by 2025, IoT applications will have an economic benefit of $3.9 to $11.1 trillion; up from $0.3–$0.9 trillion in 2015.

© Springer International Publishing AG 2017
I. Podnar Žarko et al. (Eds.): InterOSS-IoT 2016, LNCS 10218, pp. 91–106, 2017.
DOI: 10.1007/978-3-319-56877-5_6

Those studies are encouraging, since they suggest a tremendous impact of the IoT over the coming years. Nevertheless, the McKinsey analysis [3] also points out a significant threat to the estimated economic benefit: *missing interoperability*. Specifically, the authors state that a 40% share of the estimated value directly depends on interoperability between IoT systems, i.e., it can only be achieved if two or more systems are able to work together. E.g., an adaptive traffic control system of a city has more value, the more information systems it can interact with. Only if it can interoperate with different systems, e.g., for digital traffic signage, traffic lights, parking systems, or public transport, a traffic control system can reach its full potential.

Establishing interoperability on the IoT is the vision of the BIG IoT project[1] [4]. In order to support the development of cross-platform and even cross-domain applications and the emergence of entire IoT ecosystems, BIG IoT delivers key technological enablers. First, a common API among IoT platforms is developed so that application development is facilitated. Second, a marketplace as a center piece of an IoT ecosystem is introduced and implemented. The marketplace is key for enabling all stakeholders of the ecosystem to participate in revenue streams.

However, to make such interoperable IoT ecosystems possible, the benefits for all stakeholders need to be understood and pointed out. While the value for the user (e.g., a city administration) is clear, some stakeholders have protected assets and benefitting from an interoperable ecosystem is not obvious. Thus, this article studies the research question of how the different stakeholders of an interoperable IoT ecosystem can benefit and create value. Therefore, the goal of this paper is to outline the characteristics of an interoperable IoT ecosystem, identifying the relevant stakeholder roles, and analyzing potential business models. We are conducting this study as part of the BIG IoT project, with several industrial and research partners involved.

The remainder of this paper is structured as follows. Section 2 gives an overview of existing studies and related work in this field of research. Section 3 describes the key characteristics of interoperable IoT ecosystems, its stakeholders, and their relationships. In Sect. 4, we analyse and discuss potential business models for the identified stakeholders. Finally, Sect. 5 draws conclusions from our findings and points at future work in this field of research.

2   Background and Related Work

In this section, we provide an overview of research on business models for IoT ecosystems.

A very comprehensive study on the IoT market as a whole and its development can be found in [3]. Based on a view of nine vertical markets, as similarly seen in [5], a market prognosis is presented. The key findings support our goal of enabling interoperable IoT ecosystems: The authors estimate that the potential economic impact of IoT applications in nine vertical markets may be as high as $11.1 trillion per year in 2025. However, interoperability between IoT systems is critical in order to reach this impact,

---

[1] http://big-iot.eu.

and the authors expect that 40–60% of potential value is generated through cross-platform IoT applications. Further, the authors identify most sensor-collected data is currently unused, e.g., an oil rig with 30,000 sensors is examined on which only 1% of the data is being used. Also in such cases, interoperability and facilitated access to the data will help in the future to improve this ratio of data being used.

In [6], two main classes of business models are distinguished. First, *Digitally Charged Products*, which refer to the new possibilities of the digital transformation for manufacturing industries. Second, the *Sensor as a Service* idea, where sensor data are collected, processed and sold. The second group characterizes also the approach of interoperable IoT ecosystems followed by BIG IoT (see Sect. 3), where IoT data sources are offered by IoT service providers. The St. Gallen business model navigator [7] analyses 250 business models applied in the past 25 years and identifies 55 patterns being used as basis for innovation of business models in the IoT. The UNIFY project analyses in [8] a broad range of business models to provide a basis for the dialogue of the European Platforms Initiative[2] (IoT-EPI). The framework captures the challenges of building IoT ecosystem business models considering the heterogeneity of smart node devices at the edge, network technologies, multiple standardisation initiatives, the immaturity of innovation, and the unstructured ecosystems.

Following the above findings we have to distinguish between business models that (1) target end-users of the IoT and (2) those focusing on business to business revenues. The first case includes, e.g., production companies which are digitally upgrading their businesses from product selling to selling services. The second case includes business models which benefit from ecosystems and require centralized marketplaces for services and/or applications. Further, as the IoT combines the physical with the digital world and fosters cooperation between partners from different domains, a huge number of stakeholders with a wide variety of interests are involved. This makes it difficult to overview the wide variety of business models, which can be complex. So in contrast to the so far usual value chains, the more powerful tool of value networks will be useful to identify more complex relationships of participants of the ecosystem (see Sect. 3.2).

A conclusion of our related work analysis is that most of the current work is focusing on analyzing business models for device manufacturers. Analyses for IoT ecosystem value propositions are currently missing. At this point, our paper extends the current state of art by identifying the relevant stakeholders and their potential business models within an interoperable IoT ecosystem.

3 Characteristics of an Interoperable IoT Ecosystem

This section describes the need for interoperability in order to ignite IoT ecosystems and presents the BIG IoT approach (Sect. 3.1). Further, we identify different stakeholders and their relationships within such an ecosystem (Sect. 3.2), in order to derive relevant business models for those stakeholders in Sect. 4.

---

[2] http://iot-epi.eu/.

3.1 Realizing an Interoperable IoT Ecosystem – the BIG IoT Approach

The fundament of an IoT ecosystem is the "thing", i.e., physical entity with a virtual counterpart that computes/communicates information and may be controllable autonomously or remotely. These things may be directly connected and accessible through the Internet, e.g., a Raspberry Pi or smart phone, which we call a *device-level platform*. They may also be connected through a gateway, which we call a *fog-level platform*, or there is an aggregating *cloud-level platform*, which is deployed on a server [4]. A few prominent examples of cloud-level platforms are ThingWorx[3], AWS IoT[4], or Xively[5]. There are more than 360 IoT platforms today and the number is continuing to grow [9]. However, the landscape is complex; each IoT platform defines its own interface, data formats, and semantics. This situation with respect to cloud level platforms is illustrated in Fig. 1, which shows the variety of platform interfaces in form of varying shapes on the interface connector.

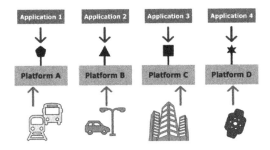

Fig. 1. The problem of missing interoperability. (Icons by Freepik from http://www.flaticon.com)

On the one hand, this situation is due to the unavailability of well-adopted standards and shared semantic vocabularies. While work on various IoT standards is in progress (e.g., oneM2M [10] or OMA LWM2M [11]), none of the more high-level standards has reached broad acceptance, yet. On the other hand, the providers of IoT platforms intentionally choose proprietary interfaces. This helps to protect their environment. Once customers have invested in applications using the proprietary interface, the platform has defensible advantages. While this may be an advantage for platform providers once they reach a large customer base, this is a disadvantage for thing providers as well as application developers. The interface heterogeneity makes cross-platform applications more difficult to realize since supporting variety of interfaces is costly and increases time to market. Especially, small enterprises cannot afford providing solutions on all different platforms, as they can only provide applications for a small number of platforms, e.g., a traffic information application for one specific city. For thing providers, e.g., the public transport organization of a city, a vendor-lock is disadvantageous as it may develop higher contracting costs in the long-run.

---

[3] https://www.thingworx.com.

[4] https://aws.amazon.com/iot.

[5] http://www.xively.com.

Today, IoT solutions are often in vertical silos with no or little interoperability between them. The BIG IoT project addresses this gap of interoperability between IoT platforms as illustrated in Fig. 2. By establishing a common API (visualized as round interface connector), called the *BIG IoT API*, services and applications can easily access different IoT platforms. Thus, in addition to existing proprietary interfaces, platform providers can support the BIG IoT API to take part in the IoT ecosystem. The common place to discover *offerings* of platforms and services for business users is the BIG IoT marketplace. Additionally, the marketplace offers all stakeholders in the ecosystem the means to trade their offerings. Offerings encompass a set of related information (e.g., low-level sensor data or aggregated information) or functions (e.g., actuation tasks or computational functions). As depicted in Fig. 2, we distinguish between services and applications. While the latter only consume offerings, services consume *and* provide offerings.

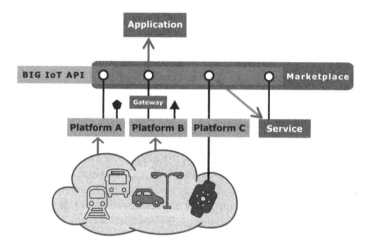

Fig. 2. BIG IoT approach towards an interoperable IoT ecosystem. (Icons by Freepik from http://www.flaticon.com)

In this way, platform providers may reach business partners who are otherwise out of reach.

3.2   The Stakeholders of an Interoperable IoT Ecosystem

In order to better understand the different stakeholders and their motivation in such an IoT ecosystem, as being realized by BIG IoT, we have created a value network model depicted in Fig. 3. Value network analysis is a business modeling methodology that visualizes business activities and sets of relationships from a dynamic whole systems perspective [12]. The nodes in this network represent different stakeholders of the IoT ecosystem. The lines between different nodes are the relationships between the stakeholders. All tangible and intangible value objects that are exchanged between different stakeholders are marked on the corresponding relationships.

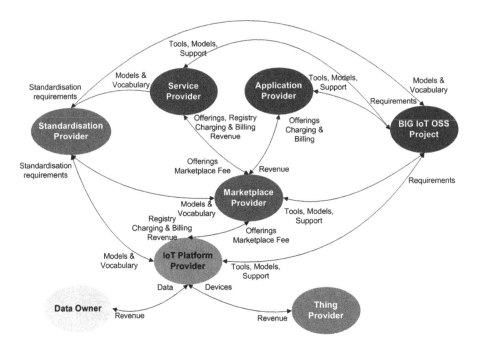

Fig. 3. Value network model for interoperable IoT ecosystems.

The *BIG IoT Open Source Software (OSS) project* provides tools, models and support to *Service* and *Application Providers*, *IoT Platform* and *Marketplace Providers* in order to enable them to use the BIG IoT technology to develop their assets. On the other side, these stakeholders provide requirements for further development of the BIG IoT OSS project. The *Thing Provider* operates or sells devices (e.g., sensors or actors) as well as objects equipped with such devices (e.g., traffic lights equipped with radar sensors). He enables the connection of the provided things to an IoT platform. The *IoT Platform Provider* has relations with the *Data Owners* whose data are being collected and provided as offerings on the marketplace. The *Marketplace Provider* on the one side facilitates the trading of offerings by providing means for offering's registration and search, as well as billing and charging for the usage of these offerings in return for the marketplace fee. On the other side, he enables the *Service Provider* to easily discover already registered offerings, build new services and then provide service output as new offerings on the marketplace, in return for the marketplace fee. *Application Providers* use the offerings traded on the marketplace to develop applications for their end customers but do not offer them on the marketplace for business customers. Last but not least, *Standardisation Providers* contribute mainly with models and vocabularies to enable semantic interoperability.

The role of the main stakeholders is studied in more details in Sect. 4.

4 Business Model Analyses

As discussed in Sect. 2, interoperability is needed to exploit the economic impact and all business opportunities emerging from the IoT. In this section, we analyse how the different stakeholders identified in Sect. 3 can enhance value propositions of their current business models in such interoperable IoT ecosystems. Further, we discuss these business models and identify the importance of a marketplace as a central point of exposition and trading of offerings from heterogeneous IoT platforms and services.

4.1 Business Model Canvases for IoT Ecosystem Stakeholders

For analyzing business models of the different stakeholders of an interoperable IoT ecosystem, we have used the established *business model canvas* methodology with its nine building blocks [13].

The business models for each stakeholder of the ecosystem are carried out taking into account those following nine steps of the business model canvas:

Step 1: Customer Segments: Describing the addressed most important customer segments and roles, for which the value is created.

Step 2: Value Propositions: Definition of the main products and services, that are delivered to the customers and creates value for them.

Step 3: Channels: Outlines the channels through which the customers are reached and served.

Step 4: Customer Relationships: Gives a description of the relationship with the customers.

Step 5: Revenue Streams: Defining the benefits and earnings for the business model for the value propositions consumed by the customers.

Step 6: Key Resources: Describing the physical, intellectual financial and human resources needed to run the business.

Step 7: Key Activities: Showing the most important activities to be performed.

Step 8: Key Partnerships: Outlining the main partners and suppliers to provide additional external activities and resources.

Step 9: Cost Structure: Defining the principal cost to setup and run the business.

In the following, the four business model canvases of the *IoT Platform Provider*, the *Service & Application Provider*, the *Thing Provider*, as well as the *Marketplace Provider* will be described. The inputs for the different building blocks have been assessed according to a survey among the industrial and research partners of the BIG IoT project and also taken from other research and productive ecosystem evaluations and examples.

Business Model Canvas of an IoT Platform Provider

By using the business model canvas (Table 1), we analyse the main opportunities for the IoT platform provider that emerge from the integration with the BIG IoT API and participating in the Marketplace.

Table 1. Business model canvas of an IoT platform provider.

Key Partners	Key Activities	Value Proposition	Customer Relationship	Customer segments
IT Vendors	Development	Data provision (domain independent)	Consultancy	Public administrations
IoT Platform Vendors	Integration		Self Service	Public utilities
Thing Providers	Operation	Data discovery	Personalized Support	SMEs
Data Owners	Sales	Reuse of data and composition		Users of IoT Data
Standardisation Providers	**Key Resources**	Services for charging and billing	**Channels**	
BIG IoT OSS project	Developers		Web	
	Data Centers	Flexible deployment model	Sales	
Marketplace Provider	Networking	Operational support	References	
			Conferences	

Cost Structure	Revenue Stream
Development	Flat rate
Integration	Fixed price
Operation	Consulting contracts
Marketing & Sales	
Support	

An IoT platform value grows if it catches demand both from the side of IoT data providers (e.g., things providers or data owners) and from the side of data users (application/service providers). The main partners of the IoT platform provider are its suppliers (i.e., IT and IoT platform vendors). As the key asset of the IoT platform provider is the content available on the platform, the range of key partners further comprises things providers, marketplace provider, and data owners. In order to take part in the ecosystem, the BIG IoT OSS project as well as standardisation bodies are becoming partners to the platform provider, since it can interact with them in order to influence interface definitions.

The core activities of the platform provider are operation on data (their exposure), development of platform services, and sale of those services. To do this, the IoT platform provider exploits storage and computing resources, developing capability, data models, and networking. The key value proposition is strictly linked with exchange and exposure of data, data combination, and operational support. Customer relationships of the IoT platform provider are often strengthened through consultancy and personal assistance devoted to customer segments, such as IoT data users (e.g., service or application providers) and IoT data producers (e.g., public administrations, or utilities). Also, small and medium sized enterprises (SMEs) are often relying as customers on IoT platform providers, as they do not have the capacity to run their own IoT platform. The main costs are derived from the development, management, and evolution of the IT infrastructure as well as the data maintenance. The IoT platform provider can expect revenue streams from the customers through recurring fees (flat rate model) or through fixed prices based on individual contracts. Also, consulting contracts, e.g., for customizing the platform to specific needs, are possible.

By participating in an ecosystem, such as the one realized by BIG IoT, the traditional business model of the IoT platform provider is strengthened, as the IoT platform becomes a product offered through the marketplace connected with the BIG IoT API. Through this registration on the marketplace, the visibility of the platform increases. The key value offered, the access and use of data, is facilitated by relying on a common API. This adds value for the customers and IoT platform users. The above advantages will eventually increase revenue streams.

The BIG IoT solution increases the benefit in following aspects of the Business Model:

Value Proposition:

- Usage of the BIG IoT API and a common offering representation in the BIG IoT marketplace enables domain independent provision of data.
- Discovery of data is made easy by the marketplace search and discovery features.
- Reuse of data and composition of data and services is enabled by semantic description of offerings, based on shared models.
- Services for charging and billing offered by the BIG IoT marketplace facilitate the platform provider to offer a common billing interface to all applications or services.
- Flexible deployment model allows working across all inter- and intra-segment partners of BIG IoT.
- Operational support can be provided with protection of investment and reusability for more than one platform.

Key Partners:

- Partnerships with different standardisation bodies and BIG IoT OSS project will ensure wider acceptance and thus higher protection of investment and future reusability of the platform interfaces.
- Marketplace Provider enables the provision of the platform offerings to a broader market than today, since currently only the segments of direct customers of the vendor and in a given vertical segment are served.

Revenue Stream:

- price for flat rate, fixed price and consulting contracts include increased value for a given price, as the same platform will be able to interoperate with more services and applications also from different domains.

Business Model Canvas of an IoT Things Provider

"Things" (the real-world objects connected to the IoT) represent the front-end of what the consumer will see, touch and feel when he first interacts with IoT technology. The device's task is to provide functionality and on a second level to interact with other connected objects in order to enhance the capabilities of an ecosystem and creating more comprehensive scenarios. Things can also generate data, which can be used by other devices or services to better accomplish their tasks. Putting these considerations in the context of BIG IoT, "things" theirselves can become *first-class citizens* of the larger IoT ecosystem, through equipping them with the commonly defined APIs. By doing so, the business model canvas in Table 2 sketches out the relevant factors of a things provider from a business perspective.

Table 2. Business model canvas of an IoT things provider.

Key Partners	Key Activities	Value Proposition	Customer Relationship	Customer segments
Module / IC providers BIG IoT OSS project	Development Integration	Provisioning of things Enabling connection of things to platforms	Support in utilisation of things	Public administrations Public utilities
	Key Resources Developers	Publishing data through common API	**Channels** Sales representatives Marketing channels	SMEs IoT Platform Providers

Cost Structure	Revenue Stream
R&D Development Operation Sales & marketing	Fixed price per unit (if things are sold) Operation contracts (if things are operated for third party) Support / service contracts

Apart from providing the things, the value proposition of the thing provider is to facilitate the connection of the things with IoT platforms. This process is supported through common APIs, such as the BIG IoT API. Additionally, the common API can mask hardware complexity and abstract from the challenges of the underlying hardware by exporting a comprehensive and common interface. Among the ecosystem partners of the thing provider are module and integrated circuit (IC) manufacturers, who provide the components on which the design of the product is based, as well as the BIG IoT OSS project, which offers software that can be reused to integrate things. Key resources to be invested are developers that realize the hard- and software. They implement the API as well as device-level applications and ensure that the process of development is smooth. Once a common and open API is chosen, the audience of developers can be extended by externals, which results in overall benefits for the ecosystem. The main cost drivers are R&D, operation, sales and marketing. The revenue stream is either coming from the operator of the things (in case thing provider sells things) or is coming from operation contracts, in case the thing provider is in charge of operating. Additionally, contracts to support the utilisation of things may generate revenue. A model that will presumably become more and more important in the future is the generation of revenue through offering things as a service (e.g., railway companies may acquire entire locomotives on service basis, i.e., they pay the things provider per day of operation). Such service model contracts are further supported through common APIs, as defined by BIG IoT.

Business Model Canvas of a Service/Application Provider

The service and application providers have a crucial role in an IoT ecosystem, as they bring additional value on top of the IoT platforms. Table 3 outlines business model considerations from their perspective.

Table 3. Business model canvas of a service/application provider.

Key Partners	Key Activities	Value Proposition	Customer Relationship	Customer segments
IoT Platform Providers	Development	Higher value information	Support	Application Providers (using a service)
Standardisation Providers	Operation	Added-value functionalities	Consulting	Service Providers (using a service)
BIG IoT OSS Project	**Key Resources**	Enrichment through value-chain	**Channels**	Business users (e.g., an organization using an application)
Marketplace Provider	Developers		Web or direct marketing	
	Marketing & sales	Common API facilitates integration		

Cost Structure	Revenue Stream
Development	Pay per use
Operation	Pay per install (in case of applications)
Marketing & sales	Support / service contracts

The service provider as well as the application provider offer a number of value propositions within an interoperable IoT ecosystem. Based on lower-level input (i.e., an IoT platform or another service), a service or application can offer either higher value information (e.g., weather forecast based on temperature, humidity, and wind measurements) or added-value functionalities (e.g., switching light off in entire building based on single light switches). This enrichment through the chaining of offerings from different parties is valuable for customers. By utilizing the common API or even exposing it (in case of services), the integration with other components of the IoT ecosystem becomes easier. Hence, customers are again other application- or service providers with high-level capabilities, or also business users, e.g., organizations which utilize an application. Relationships to these customers can be maintained through support or even specific consulting. These activities are also a possible revenue stream, apart from the pay per use or a direct payment for the service / application. The key partners of the service and application provider are IoT platform providers, marketplace provider, BIG IoT OSS project, standardisation providers as well as developers. The main activities are development, operation and marketing.

Business Model Canvas of a Marketplace Provider

In the previous canvases we presented how the different stakeholders can enrich their value proposition to their customers by participating in an IoT ecosystem, e.g., through the BIG IoT solution. The following business model canvas (Table 4) summarizes the value proposition of the Marketplace to these stakeholders.

Table 4. Business model canvas of marketplace provider.

Key Partners	Key Activities	Value Proposition	Customer Relationship	Customer segments
IoT Platform Providers	Development	Discovery of offerings	Support	Service Providers
Standardisation Providers	Operation	Advertisement of offerings and broadening of customer outreach	Consulting	IoT Platform Providers
BIG IoT OSS Project	Product Management			Application Providers
	Key Resources	Charging and billing	**Channels**	
Developers	Developers	Management of common vocabulary	Web and direct marketing	
IT vendors	Marketing & sales		Platform, service, and application vendors	

Cost Structure	Revenue Stream
Development	Advertising fees
Operation & infrastructure	Pay per use
Support	Percentage of each payment
Traffic generation and retention	Entry fees
	Support / consulting contracts

The key value proposition of the marketplace is enabling the discovery of offerings from IoT platforms or value adding services. This discovery is provided as searching capabilities on a user interface, as well as through a machine readable API. Applications are specifically not listed in the marketplace of BIG IoT, as there are already many established app stores for this purpose. Nevertheless, also application providers (besides service- and platform providers) are the main customers of the marketplace. All stakeholders profit from the advertisement (or: "marketing") capabilities of the marketplace, which broadens the customer outreach of those offering providers. The discovery and advertisement of offerings is supported through the management of common vocabularies by the marketplace. This is the key to semantic interoperability within an IoT ecosystem. Common terms (e.g., "traffic light" or "temperature"), which are used by multiple participants of the ecosystem, are registered and referenced at the marketplace. Beyond these capabilities for reaching interoperability, the marketplace supports charging and billing. I.e., a service or platform can state how much access to their offerings costs and consumers of those offerings have to pay. Through these functionalities, the marketplace enables the monetization of IoT offerings.

To operate a marketplace, its provider mainly invests into development and operation, but also product management (i.e., marketing, feedback, promotion, sales) is a key activity for success. Thereby, customer relationships can be initiated through consultancy, customizing assistance, and support. Then, revenue streams will be generated through contractual work for those activities. Apart from those, the marketplace has several interesting possibilities for creating revenue based on different payment models. These range from fees for better advertisement, over a pay per use (e.g., counting API calls), small participations in each payment, up to entry fees for service and platform providers to enlist their offerings.

4.2 Discussion on IoT Ecosystem Business Models

The analyses above show that for each stakeholder, business models can be identified within an interoperable IoT ecosystem. From our perspective, all stakeholders can profit from interoperability and the creation of an IoT ecosystem. Naturally, their effectiveness can only be evaluated in practice. However, the success of an IoT ecosystem will depend (a) on the willingness of IoT platform providers and platform vendors to adopt common APIs into their platforms so that a sufficient offer of data is available and (b) on the number of service and application providers to use these and add value to the data via their offerings. I.e., the lower the initial barriers to enter the ecosystem and a marketplace, the more likely will be the success.

Once a marketplace is established, IoT offerings of platforms and services can be easier found and used to create new, innovative applications. By means of semantic search of offerings, service- and application providers can find resources from different platforms and domains that best fit their needs. Additionally, by using a common API and vocabularies a service provider can more easily provide and trade its offerings. In this way, they can more rapidly deliver services to their existing customers and reach new customers. Furthermore, by using charging and billing of the marketplace they can outsource these activities.

As discussed in Sect. 3, value chains are evolving towards a value network comprising multiple stakeholders in the ecosystem. When taking the primary functionality of providing a marketplace for the offerings, a general view on the clients of the marketplace only distinguishes between offering providers and offering consumers, as shown in Fig. 4.

By bringing together the offering providers (platforms and service providers) and the offering consumers (services and application providers) the marketplace fosters the exploitation across the complete value network of an IoT ecosystem. The marketplace even pushes the utilisation for all involved ecosystem stakeholders due to interoperable APIs and the advanced discovery as well as monetization facilities.

To evaluate from an application/industry point of view the value and benefit, we have to investigate in the future through the lens of individual industries or sectors (see [3, 14]). The existing vertical customer segments of whole industries will be affected by enhancement of IoT capabilities. They will cover more or less all market sectors, but with respect to IoT some will gain more potential than others. In particular, the following vertical markets are important for the IoT [3]: Factories, Cities, Retail environments,

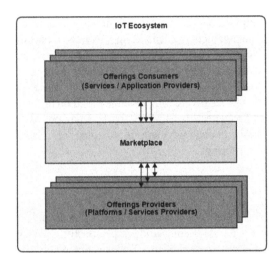

Fig. 4. The marketplace as centre piece of an interoperable IoT ecosystem.

Work sites, Vehicles, Agriculture, Outside, Home, Offices. The interoperability and marketplace create value for business users across settings and sectors. As a marketplace can provide presentation and promotion of the offerings relevant across multiple vertical segments as well as semantic search options, the ecosystem is stimulated to push inter-segment and intra-segment value generation as illustrated in Fig. 5.

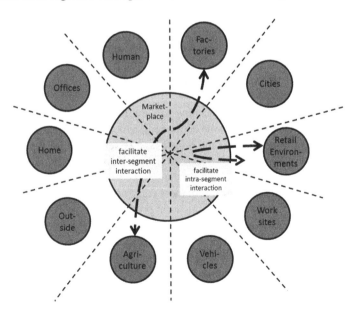

Fig. 5. Marketplace facilitates inter-segment and intra-segment interaction.

5   Conclusions and Outlook

In this paper we present an overview of the IoT ecosystem and its stakeholders and the advantages interoperability can bring for them. Starting from a description of the BIG IoT solution, as a realisation of an IoT ecosystem, we argue that interoperability brings new business opportunities for all participants in such an ecosystem. By using the value network model analysis we identify the key stakeholders, relationships, as well as tangible and intangible value exchange between different roles. Further, based on the business model canvas method, we analyse existing business models of four key stakeholders and identify how these models are being enhanced through an interoperable IoT ecosystem to provide more value to their customers. In our discussion, we identify the marketplace as the fulcrum of such an ecosystem, and we explain the importance of this role for the inter-segment and intra-segment interaction.

In the future, we will further study the final designs of revenue schemes and which business models are most suitable for the economic success. This work will be done alongside the implementation of three different pilots of the BIG IoT project in Barcelona, Berlin/Wolfsburg, as well as Piedmont. Furthermore, we will investigate how orchestration between all kinds of IoT services and offerings can be supported through the marketplace. Automated orchestration promises to reduce costs through less adaption efforts and empowerment of IoT end-to-end use cases.

Acknowledgments.   This work is financially supported by the project "Bridging the Interoperability Gap" (BIG IoT) funded by the European Commission's Horizon 2020 research and innovation program under grant agreement No. 688038.

References

1. Gershenfeld, N., Krikorian, R., Cohen, D.: The Internet of Things. Sci. Am. **291**, 76–81 (2004)
2. Gartner: Gartner says the Internet of Things installed base will grow to 26 Billion units by 2020 (2013). http://www.gartner.com/newsroom/id/2636073
3. Manyika, J., Chui, M., Bisson, P., Woetzel, J., Dobbs, R., Bughin, J., Aharon, D.: The Internet of Things: Mapping the Value Beyond the Hype. McKinsey Global Institute (2015)
4. Bröring, A., Schmid, S., Schindhelm, C.-K., Khelil, A., Kaebisch, S., Kramer, D., Le Phuoc, D., Mitic, J., Anicic, D., Teniente, E.: Enabling IoT ecosystems through platform interoperability. IEEE Softw. **34**(1), 54–61 (2017)
5. James, R.: The internet of things: a study in hype, reality, disruption, and Growth. Raymond James US Research, Technology & Communications, Industry Report (2014)
6. Fleisch, E., Weinberger, M., Wortmann, F.: Business models and the internet of things. In: Interoperability and Open-Source Solutions for the Internet of Things, pp. 6–10. Springer (2015)
7. Gassmann, O., Karolin, F., Michaela, C.: The St. Gallen business model navigator. ITEM-HSG, St. Gallen (2013)
8. Vermesan, O., Bahr, R.,Gluhak, A., Boesenberg, F., Hoeer, A., Osella, M.: IoT business models framework. Unify-IoT Project (2016)
9. List Of 360 + IoT Platform Companies. IoT Analytics (2016). https://iot-analytics.com/product/list-of-360-iot-platform-companies/

10. Swetina, J., Lu, G., Jacobs, P., Ennesser, F., Song, J.: Toward a standardized common M2M service layer platform: Introduction to oneM2M. IEEE Wirel. Commun. **21**, 20–26 (2014)
11. Open Mobile Alliance. Lightweight Machine to Machine Technical Specification, Candidate (2015)
12. Allee, V.: What is ValueNet works analysis?. In: ValueNet Works Fieldbook (2006)
13. Osterwalder, A., Pigneur, Y., Bernarda, G., Smith, A.: Value Proposition Design: How to Create Products and Services Customers Want. Wiley, New York (2014)
14. IoT market segments – biggest opportunities in industrial manufacturing. IoT Analytics (2016). https://iot-analytics.com/iot-market-segments-analysis/

On the Road to Secure and Privacy-Preserving IoT Ecosystems

Juan Hernández-Serrano[1]([✉]), Jose L. Muñoz[1], Arne Bröring[2], Oscar Esparza[1], Lars Mikkelsen[3], Wolfgang Schwarzott[4], Olga León[1], and Jan Zibuschka[5]

[1] Universitat Politècnica de Catalunya, Barcelona, Spain
jserrano@entel.upc.edu
[2] Siemens AG, Munich, Germany
[3] Aalborg Universitet, Aalborg, Denmark
[4] Atos IT Solutions and Services GmbH, Vienna, Austria
[5] Robert Bosch GmbH, Stuttgart, Germany

Abstract. The Internet of Things (IoT) is on the rise. Today, various IoT platforms are already available, giving access to myriads of *things*. Initiatives such as *BIG IoT* are bringing those IoT platforms together in order to form ecosystems. BIG IoT aims to facilitate cross-platform and cross-domain application developments and establish centralized marketplaces to allow resource monetization. This combination of multi-platform applications, heterogeneity of the IoT, as well as enabling marketing and accounting of resources results in crucial challenges for security and privacy. Hence, this article analyses the requirements for security in IoT ecosystems and outlines solutions followed in the BIG IoT project to tackle those challenges. Concrete analysis of an IoT use case covering aspects such as public, private transportation, and smart parking is also presented.

Keywords: Internet of Things · IoT · Security · Privacy

1 Introduction

In the past years, the Internet of Things (IoT) has largely expanded and the number of IoT devices is ever increasing. Today, IoT use cases span over a wide variety of application domains, ranging from smart homes over e-health systems to industrial environments. *Things* used in such applications are made available through IoT platforms. These platforms can be located on the device, edge, fog, or cloud levels.

A multitude of such platforms exists today. In order to enable cross-platform and even cross-domain application development, different initiatives are determined to form IoT ecosystems. An example for this is BIG IoT[1] [9]. At the moment, the BIG IoT project comprises 8 IoT platforms and is ready to grow beyond them by means of an open call. To ignite such an IoT ecosystem, BIG IoT focuses on establishing interoperability across platforms.

---

[1] http://big-iot.eu.

I. Podnar Žarko et al. (Eds.): InterOSS-IoT 2016, LNCS 10218, pp. 107–122, 2017.
DOI: 10.1007/978-3-319-56877-5_7

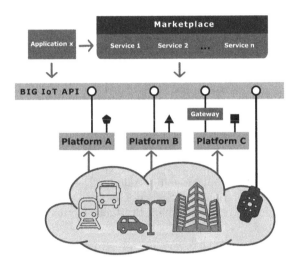

Fig. 1. The BIG IoT approach for building an ecosystem of IoT platforms. (Icons by Freepik from http://www.flaticon.com)

BIG IoT has two main objectives. The first one is defining a shared interface, i.e., the so-called BIG IoT API comprising common functionalities such as discovery, access, and event handling. This API needs to be supported by all participating platforms, often in addition to their existing proprietary interface, as illustrated in Fig. 1. The second objective is establishing a centralized marketplace where platforms as well as value-adding services can be registered, searched, and subscribed for by applications. In the BIG IoT project, these technologies are deployed in multiple pilot scenarios and involving various IoT platforms, services, and applications from the Smart Cities domain. We will provide an example of these scenarios in Sect. 4.

Besides the evident benefits that can be achieved by such IoT ecosystems, there are crucial challenges to deal with. In particular, new security threats must be addressed to allow the continued growth of such ecosystems. Frequently, sensitive data are stored, sent, or received by IoT platforms. Thus, security mechanisms are needed to protect these data from unauthorized access. Consider a patient who is wearing a glucose sensor that transmits its results to the IoT platform of a medical centre. Security vulnerabilities may allow other entities to misuse this information or even put at risk the physical safety of the patient if these data are forged.

Dealing with IoT security risks is challenging and can be more complex than in conventional networks, particularly for companies entering IoT ecosystems without any experience in the security field. Moreover, as new security vulnerabilities may be discovered over time, there is a need for updating IoT platforms on a regular basis. This might be hard to achieve in some cases either due to the simplicity of some device-level IoT platforms, or due to the lack of awareness of users or platform admins that forget or just skip updates. Finally, it may happen that some IoT platform manufacturers decide not to provide ongoing support nor security updates in order to reduce costs.

Last, but not least, privacy must be a mandatory concern. A privacy analysis should at the least find an appropriate answer to the question: do the collected data allow drawing conclusions on individual or small, specific groups or human beings? Note that such conclusions may be drawn by unauthorized eavesdroppers, and then this discussion is overlapping with confidentiality.

The purpose of this article is to outline current discussion, analysis, and specific actions with regard to security and privacy in IoT ecosystems, and particularly to the BIG IoT realization of such an ecosystem. Requirements and best practices presented here will help to secure all the assets of the BIG IoT ecosystem and to prevent abuse of sensitive data.

The rest of this document is structured as follows. Section 2 presents a set of security requirements for the BIG IoT as well as current discussion and actions in order to address them. Section 3 outlines privacy recommendations for the IoT ecosystem. Next, in Sect. 4 a BIG IoT use case example is presented and analysed from the point of view of security and privacy. Finally, Sect. 5 provides with the conclusions of this work.

2   Securing an IoT Ecosystem

The BIG IoT marketplace, the common API, and all the platforms/services/ applications in the ecosystem must comply with a set of security requirements. After an analysis of the BIG IoT needs, seven security requirements were identified, which are presented in Sect. 2.1. Moreover, in order to face these risks, some solutions were already discussed (see Sect. 2.2).

2.1   Requirements

1. **End-to-end security**. IoT communications typically spread over several nodes and technologies. In particular, BIG IoT is not another IoT platform, it is a framework for a heterogeneous set of platforms, services, and applications. A possible solution to provide security would be to leave the mechanisms already in use for each platform, and then to define adaptation policies of these mechanisms in the boundary points of platforms. The definition of these "low-level" relationships would highly increase complexity, as each individual security protocol (suite) provided by a component would have to be mapped to each protocol (suite) offered by each component it communicates with, which may fail, and hence should be avoided. The solution adopted in BIG IoT is to provide security at the API level, because it is common for all platforms. So, there is no need to adapt protection mechanisms between platforms, as the API is end-to-end by nature and assures that security remains independent of low level platform components.

2. **"Batteries included but swappable"**[2]. BIG IoT has to be designed to be capable of *ageing* in place while still addressing evolving risks [22]. There may appear new attacks, crypto systems, counter measures, techniques, and topologies, but the IoT system must be capable of dealing with these emerging concerns long after the system was deployed. Consequently, BIG IoT must ship a default but swappable security implementation, not hard-coded to specific security protocols/systems. Notice that this is not incompatible with defining current minimum requirements in terms of protocols, mechanisms and/or algorithms.

3. **Flexible authentication/authorization**. The authentication and authorization systems used in the BIG IoT ecosystem must ease the management of identities and permissions. The various platform that BIG IoT composes for interoperation have very different user management architectures and interfaces. Furthermore, platform operators may not want to expose certain user management features due to business or even security concerns. Therefore, supporting various models including decentralized, federated or delegated authentication is required for successful interoperation of platforms' user management systems.

4. **Ownership transfer**. BIG IoT should support safe transfer of ownership, even if a component is sold or transferred to a competitor; something that often happens during the lifespan of IoT nodes/components.

5. **Accounting and charging**. The BIG IoT must implement a secure accounting of resources consumption. This accounting must generate enough charging data, typically in the form of a Charging Data Record (CDR), so that the desired charging policies can be enforced. As a result of a charging policy, a billing system may be necessary to generate invoices for service consumers. All these systems must be flexible enough to implement different business models and monetization strategies of services that can be implemented in the BIG IoT ecosystem. The BIG IoT marketplace must support offline and online charging and billing.

6. **Continuous security**. The BIG IoT system should be ready to respond to hostile participants, compromised nodes, and any other adverse event. Therefore, it is necessary to implement mechanisms and/or tools to re-issue credentials, exclude participants, distribute security patches, updates, swap algorithms, or protocols, etc.

7. **Secure development**. Security must be a key part during the design phase of every BIG IoT software, but a secure design would be useless if development errors open unexpected attacks and/or vulnerabilities. Using a Secure Software Development Life Cycle (S-SDLC) and secure Source Code Analysis (SCA) would help developers to build more secure software and address security compliance requirements.

---

[2] https://blog.docker.com/2016/03/docker-networking-design-philosophy/.

2.2   Addressing the Security Requirements in BIG IoT

Even though many strategies or decisions are still to be taken, some actions have already been adopted in order to address the above requirements.

Requirement 1 is directly met as the BIG IoT API is an HTTP(s) based API, and so it is end-to-end by design. Moreover, in order to comply with *Requirement 2*, the API should be flexible enough to handle any protocol and/or content. BIG IoT handles this by defining a very generic API; semantic annotations of the syntactic descriptions of each registered service and platform are then used to clarify the details on how to establish communication with these components.

Requirement 3 states that there is a need of providing flexible authentication in the IoT ecosystem. I.e., BIG IoT must implement an authentication and authorization system to be shared by participating platforms, services, applications, and end-users. Moreover, BIG IoT has to be able to work even when the authentication managers are not available. To solve this, BIG IoT uses an approach that is similar to the ones used by other widely-known IoT initiatives (e.g., [6]): signed manifests or tokens. A client presents a signed manifest to a server to demonstrate that it is able to perform a given action on a given asset. When the server receives the signed manifest, it can trust the contents because the manifest is signed by a common root of trust.

Many state-of-the-art technologies have already dealt with the fact of using such signed manifests. Most solutions for the Web use JSON, CBOR, or XML encodings and rely on JSON Web Encryption (JWE) [17], JSON Web Signature (JWS) [16], XML Encryption (XML-Enc) [15], or XML Signature (XML-Sig) [7]. Obviously, one can decide to design a custom solution from scratch, which may seem at-a-glance a better suited solution. However, experience tells us that security protocols are subtle and often tricky. Consequently, BIG IoT position is to adopt existing, already tested, security technologies. The specific set of solutions is still to be decided though. Given that the BIG IoT API relies on HTTP REST, potential candidates are SAML [20], Oauth 1 [13], OAuth 2 [14], or OpenID Connect [19] (built upon OAuth 2), supporting delegated authorization and authentication/identification.

Requirement 4 must also be considered in the choice of the previous bottom-technology. The authentication/authorization system has to be defined with focus on easy management of identities and permissions, easing actions that are quite common in the IoT. This includes safe transfer of resources' ownership and quick response to dynamic topologies with frequent admissions and withdrawals.

Requirement 5 states that an appropriate accounting is key to develop charging/billing systems, both offline or online. An offline charging system just stores a CDR containing the relevant accounting and charging information (starting and ending time, data used, bandwidth, etc.). Then, the user is charged after resources have been used. In general, users being charged offline provide a bank account to pay the corresponding bill. On the contrary, when using online charging, the user typically buys a prepaid amount of credit. In this case, the charging system has to monitor online the resources consumption and then, needs to stop (or constrain) the service when the credit limit is reached. In both approaches (offline and online), it should be desirable to have non-repudiation proofs for both, the users and the marketplace to be able to verify consumptions, bills, etc. and to solve possible inconsistencies.

Requirement 6 forces the marketplace to host a secure repository where to securely download software and software updates/patches. This is a challenge that has often been addressed in the past and present. Experience here says that, apart from security, success depends on the ease of use for both end users and developers. The app stores of Apple, Google and Amazon are good examples, but BIG Iot is aiming for a more open approach for this component.

Requirement 7 requires the use of S-SDLC. To accomplish this, BIG IoT developers are following best practices for secure software development set up by the Open Web Applications Security Project (OWASP) [1]. First, the organization itself has to fulfil security related activities and software security practices, which are described in the OWASP Software Assurance Maturity Model (SAMM) [3] framework. Second, the applications meet requirements based on the OWASP Application Security Verification Standard (ASVS) [21]. Third, the application source code is to be analysed according to the OWASP secure Source Code Analysis (SCA) guidelines [2]. And finally the application will be tested for vulnerabilities and design flaws according the OWASP testing guidelines [18].

OWASP ASVS defines security requirements for applications and services. This standard currently defines 19 verification requirements. All these requirements have three security verification levels, with each level increasing in depth: ASVS Level 1 "Opportunistic" is meant for all software and its compliance adequately defends against application security vulnerabilities that are easy to discover; ASVS Level 2 "Standard" is meant for applications that contain sensitive data, such as business-to-business transactions, including those that process health-care information, implement business-critical or sensitive functions, or process other sensitive assets; and ASVS Level 3 "Advanced" is meant for the most critical applications, that is, applications that perform high value transactions, contain sensitive medical data, or any application that requires the highest level of trust. Responsibilities include controls for ensuring confidentiality (e.g. encryption), integrity (e.g. transactions, input validation), availability (e.g. handling load gracefully), authentication (including between systems), non-repudiation, authorization, and auditing (logging). Each ASVS level contains a list of security requirements, and each of these requirements can also be mapped to security-specific features and capabilities that must be built into software by developers. For BIG IoT, developers should (at least) follow the ASVS level 2 requirements, and they could complete these with level 3 requirements according to the appropriate criticality.

3   Best Practices for Privacy in IoT Ecosystems

Igniting an IoT ecosystem involves handling big data. Often these data contain sensitive information and therefore their use could be a threat to users' privacy.

The FTC published in 2015 a guide containing best practices for privacy in IoT [12] that is summarized with the following statement: while flexibility in terms of data gathering is key to innovate around new uses of data, the amount of data storage should be balanced with the interests in limiting the privacy and data security risks to consumers.

These recommendations are useful and valid in the European scope. However, they are rather generic and they should be always complemented with a specific analysis of every use case (an example is provided in Sect. 4). In the following, we provide the main ideas behind the FTC recommendations.

3.1   Data Minimisation

Data minimisation is a long-standing principle of privacy protection [10] that means that a data controller[3] should limit the collection of personal information to what is directly relevant and necessary to accomplish a specific purpose. Since users' privacy is (or it should be) key for a wide adoption of the IoT, data minimisation is key to fostering the IoT ecosystem. Indeed, data minimisation can help guard against two privacy-related risks.

First, storing huge volumes of data increases the likelihood of receiving a data breach since the is more potential harm derived from such an event.

Second, collecting and storing large amounts of data also increases the risk of using the data in a way that departs from consumers' reasonable expectations.

To minimise these risks, organizations should develop data minimisation policies and practices providing answers to questions like what types of data it is collecting, to what end, and how long it should be stored. Such an exercise is part of a privacy-by-design approach and helps ensure that a company is sensitive with data collection practices.

In the EU, the data minimisation principle derives from Article 6.1(b) and (c) of Directive 95/46/EC [10] and Article 4.1(b) and (c) of Regulation EC 45/2001 [11], which state that personal data must be "collected for specified, explicit and legitimate purposes" and it must be "adequate, relevant and not excessive in relation to the purposes for which they are collected and/or further processed".

When a company needs to gather and store sensitive data with a business purpose, it should consider whether it can do so with a deidentified data set. Deidentified data can reduce potential consumer harm while still promoting beneficial societal uses of the information.

A key to effective deidentification (anonymisation) is to ensure that the data cannot be reasonably reidentified even with external cross sources. This usually requires removing identifiers or pseudoidentifiers. Although, at first glance it seems quite affordable, recognizing non-evident identifiers is quite a challenge that often has to be faced in a manual, use-case-specific manner.

In BIG IoT, for every specific use case, an analysis of potential identifiers among the data and/or metadata stored/exchanged is being performed. Data

---

[3] An entity processing and/or storing personal information.

minimisation is encouraged specially for the BIG IoT platforms and should account for cross data not only from other BIG IoT platform/services, but also from any other external source. An example for such data minimization technologies in the context of a BIG IoT use case is given in Sect. 4.

Notice that there is a common misconception about the added costs for data minimisation. Enhancing privacy by means of data minimisation techniques does not necessarily imply added costs. Indeed, data minimisation reduces the sensitiveness of data and hence lower security would be required. As a consequence, for instance, in BIG IoT, important saving can be obtained in development costs due to a reduced ASVS level compliance.

3.2   Strong Accountability

As aforementioned, deidentified data sets can reduce many privacy risks. However, there is always a chance that supposedly deidentified data could be reidentified; especially because of the technology advances. For this reason, companies should have accountability mechanisms in place. In this context, the FTC has stated that companies stating that they maintain deidentified or anonymous data must meet three actions: (1) take reasonable steps to deidentify data, including by keeping up with technological developments; (2) publicly commit not to reidentify the data; and (3) have enforceable contracts in place with any third parties with whom they share the data, requiring the third parties to commit not to reidentify the data. This approach ensures that if the data are reidentified in the future, regulators can hold the company responsible.

Consequently, BIG IoT platforms, services, and applications should provide proper accounting mechanisms to securely log any action by any actor dealing with sensitive data.

3.3   Transparency and Easy Access

The centrepiece legislation at EU level in the field of data protection is the "Data Protection Directive" [10] which is implemented in EU Member States through national laws. This directive aims to protect the rights and freedoms of persons with respect to the processing of personal data by laying down guidelines that determine when the processing is lawful. The guidelines mainly relate to the quality of the data, the legitimacy of the processing, the processing of special categories of data, information to be given to the data subject, the data subject's right of access to data, the right to object to the processing of data, the confidentiality and security of processing and the notification of the processing to a supervisory authority. The Directive also sets out principles for the transfer of personal data to third countries and provides for the establishment of data protection authorities in each EU Member State.

In general, the conclusion is that EU's individuals need better information on data protection policies and about what happens to their data when it is processed by online services. As a result, the EU will require European organizations to publish transparent and easily accessible data protection policies.

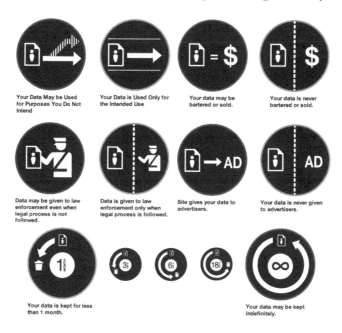

Fig. 2. Privacy icons proposal by Aza Raskin [23]

In this context, simple icons on websites and applications could explain how, by whom and under whose responsibility personal data will be processed. As a consequence, users are better informed about how and if their personal data is being exploited.

To better picture this fact, Fig. 2 shows a set of privacy icons proposed by Aza Raskin [23]. BIG IoT services and/or applications should use similar icons (or even those) to clearly show end users how their data are being processed.

4   Use Case Example: Smart Transportation Assistant

In this section we describe the use case of a transportation assistant in context of BIG IoT and analyze and discuss its security and privacy aspects.

In this case, a subscriber of the app is at home and she wants to go to a specific place. The BIG IoT app allows her to be assisted in this decision by providing information about private and public transportation.

Regarding private transportation, she can receive information about current traffic conditions. If she decides to use her private vehicle, she is assisted with navigation information while driving and is also assisted in finding available parking spots.

Regarding public transportation, the app can suggest several possible ways of transportation as an alternative to using the private vehicle. For instance, the app allows the user to select a bus line of interest. From this she can see live information about the next bus arriving at the selected stop. This information includes indications of where the bus is located currently, if the bus is delayed, the number of people on the bus, and a forecast of the number of people on the bus when it reaches her stop. Based on this information, she can choose if she takes this bus or if she should switch to take another line, a later bus, or another mode of transportation. She can also choose to see historical information about the number of people on the bus based on location and time of day. This will allow her to plan ahead, i.e. if she wants to avoid overfilled buses she can see at which times of the day the buses are less loaded.

We would like to mention that the previous use case is in fact implemented in two different pilots of the BIG IoT project. One pilot (mainly focused on private transportation) is going to be deployed in Barcelona and the other one (mainly focused in public transportation) is going to be deployed in Wolfsburg.

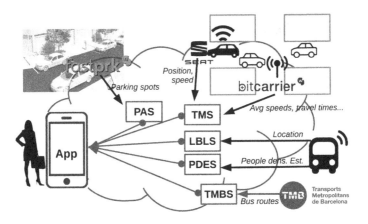

Fig. 3. Usecase: a transportation assistant in the BIG ecosystem

Finally, it is important to notice that from the users point of view, all the functionality is accessed with a single smartphone app; however, behind the scenes this app is consuming data from several services, which in turn consume data provided by physical sensors operated by different platforms. A complete picture is presented in Fig. 3 (acronyms are explained in the sections corresponding to the individual components below).

Services are an important abstraction layer in BIG IoT because they allow code re-utilization and simplify the process of building BIG IoT applications. For example, a service can aggregate data acquired from different platforms and then, present a unified dataset to apps. Another service could manage the history of data, allowing the user to access to past data (data not currently available in the source platform). Another service could create forecasts from data (acquiring data from a platform or from another service). Finally, another interesting use could be a service that anonymises an underlying dataset. This could allow lower levels of ASVS security in the upper layer (app). In the next sections we describe the involved BIG IoT components and discuss their security and privacy aspects.

Platform 1: Bitcarrier's WiFi/Bluetooth Antennas [4]

This platform is fed by data gathered by WiFi/Bluetooth antennas placed at several street crosses. The technology detects vehicles/users by their unique MAC addresses and provides average travel times, speed, and even street congestion.

MAC addresses are very sensitive data, which could be used to track or profile vehicles' (and even users') habits. This potential privacy invasion should be avoided. To do so, all the parties involved have to agree the necessary legal contracts in which they accept to properly use the data stored/exchanged. Also, these data must be appropriately secured/anonymised to avoid any kind of leakage (mistakenly or on purpose).

Under this assumption, BIG IoT approach is to immediately anonymise unique addresses using a one way cryptographic-hash function. This function uses as inputs: (1) the address to be anonymised and (2) a key. This key is updated periodically, e.g. ranging from minutes to days. In this manner, one device cannot be tracked for more than a period. How often the key is updated is part of the privacy policy.

The use of cryptographic hash functions allows anonymising the data while keeping a trapdoor that could be used to reidentify vehicles/users. The platform operator must keep secret the temporal keys used to anonymise the identifiers. However, operators may be forced to disclose these keys under some circumstances, e.g. a law enforcement requirement when a legal process is followed.

From the above reasoning, this platform should comply at least with recommended baseline ASVS level 2 "standard". In addition, the management of the anonymisation keys should comply with ASVS level 3 "advanced", as it may allow an attacker to identify/track users and/or vehicles.

Platform 2: SEAT's Cars

SEAT has put several cars with integrated sensors at BIG IoT disposal. These cars send their current position and speed. Providing these data may allow to track a vehicle and thus can be considered a threat to privacy.

Same reasoning as for the platform 1 is applied: if identifiers cannot be removed from provided data, at least they must be properly anonymised, e.g. with a cryptographic hash function. Therefore, the same ASVS security requirements apply for both Bitcarrier's platform and SEAT cars; that is, cars should comply at least with standard ASVS level 2, but the key manager (if needed) would require ASVS level 3.

Platform 3: Fastprk's On-street Parking Spot Status [5]

This platform can provide individual status of parking spots over a predefined monitored area. For instance, for the BIG IoT Barcelona use cases, this platform currently offers status information for 600 on-street parking spots. This kind of data entails specific privacy risks due to correlation with other sources: if an attacker knows where someone has parked their car, it can monitor when he/she leaves by checking the spot status.

Obviously, a straightforward countermeasure would be, e.g., to provide free spots in a given street segment (a virtual lot). This approach will guarantee k-anonymity (a given individual cannot be differentiated from another $k - 1$) of monitored vehicles/users with k being the number of vehicles parked on the same segment. This fulfils the requirement of anonymisation/deidentification of sensitive information stated before, in this case location information. The greater the segments are, the more anonymous the service is, but the PAS will provide less specific, potentially less useful, information.

Intuitively, it seems that obtaining the exact free parking spot position or the segment where there is one (or more) free parking spots is likely to be equally useful for the end user; although looking for the appropriate trade-off between privacy and usability requires further technical discussion and studies of real users' needs. While some applications/services may allow different per-user degrees of privacy, this is not the case for this scenario. Therefore, testing user feedback about the suitability (or not) of just providing free spots on the street without their specific location cannot be done on an individual basis; it should be a global approach with, e.g., a pilot project.

Since the data stored by the platform can be somehow used to track/monitor end users, we recommend securing this BIG IoT platform following ASVS verification level 2. However, if the platform just stores and provides free parking spots in a predefined segment/lot, ASVS verification level 1 could be considered.

Platform 4: Wifi Probe Catching Sensors on Buses

Wifi probe catching sensors are sensors placed on buses that collect wifi probe requests, which contain MAC addresses, emitted from users wifi enabled devices.

Since the MAC addresses are unique, they must be anonymised in the same way as it is done for platform 1. Indeed, the operation of this platform is very similar to the operation of platform 1 and therefore the same security and privacy recommendations apply.

Platform 5: Location Sensors on Bus

These sensors, placed in buses, provide location data and timestamps. The collected data is not stored at the sensor, as an outdated location would not be of much use. Since positions of public buses is not private, no specific privacy actions has to be taken. The software development has to comply with standard ASVS level 2 and, if properly justified, even with ASVS level 1.

Service 1: Traffic Monitoring Service (TMS)

This service is providing routes of cars to destinations based on current traffic conditions. With such a purpose, in this use case, the TMS consumes data provided by platforms 1 and 2 as well as a city map.

Assuming that data provided by platforms 1 and 2 is anonymised, no specific requirements in terms of privacy are required. Regarding security, the software development has to comply with standard ASVS level 2 and, if properly justified, even with ASVS level 1.

Service 2: Parking Availability Service (PAS)

In this use case, the PAS is fed with on-street parking spot status provided by platform 3. Moreover, the PAS may store statistics/historic of parking spot status. Therefore, the same privacy recommendations as for the Fastprk's platform applies here. Both the service and its connections must be protected with ASVS level 2 or ASVS level 1 if an acceptable level of k-anonymity is provided.

Service 3: External TMB Bus Routing Data Service (TMBS)

Transports Metropolitans de Barcelona (TMB) is the main public transport operator in the Barcelona metropolitan area. It already has an open data API [24] where to obtain routes to destinations with different public transports: trains, metro, buses.

The TMBS makes the TMB API in the BIG IoT ecosystem. Since all the data involved in the service are public, no specific requirements in terms of privacy are required. Regarding security, unless properly justified, the software development has to comply with standard ASVS level 2.

Service 4: People Density Estimation on Bus Service (PDES)

The PEDS consumes data from the Wifi probe catching sensors and it provides information about the number of people on buses to Public transport load application. The provided data consists of a bus id, estimation of number of people, accuracy indicator and timestamp. The provided data is stored for a fixed duration at the service, meaning that detailed load information on a specific bus can be requested within this duration. After the duration the data is minimised, such that only more general historic information is stored at the service, which is also made available to apps and services. It is recommended that the service complies with ASVS level 2, but it could even be ASVS level 1 as no user specific data is handled. However, to protect the business case of the service, i.e. to control who has access to the data and can use it, ASVS level 2 is recommended.

Service 5: Live Bus Location Service (LBLS)

This service consumes data from the location sensors on buses and provides information to Live bus location app. The provided data consists of sensor ID, location, and timestamp. The data is stored at the service for a short fixed time duration, after which it is deleted. The service does not handle any user specific data but more publicly available information, why it is recommended to comply with ASVS level 1.

Smartphone App for Enduser

An App consumes data provided by the 5 services. Apps in BIG IoT are expected to offer transparency functions, as mentioned in the privacy recommendations above, including e.g. privacy icons. Developers would also have to consider privacy requirements going beyond the scope of the BIG IoT infrastructure, such as informed consent or right of access to personal information, rectification, or deletion. Regarding secure development, apps should follow ASVS level 2.

5   Conclusions and Outlook

Nowadays, a plethora of IoT platforms and solutions exist, but yet no large-scale and cross-platform IoT ecosystems have been developed. This is mainly due to the fragmentation of IoT platforms and interfaces, as this variety results in high market entry barriers. The BIG IoT project aims at establishing interoperability across platforms in order to ignite an IoT ecosystem. Core technological pillars of BIG IoT are a common API as well as a marketplace for all participants of the IoT ecosystem, including devices, end-users, and service providers. Key to its success is to define appropriate levels of security and privacy.

Regarding security, in this paper we have identified seven requirements to be followed when creating and/or deploying BIG IoT components. Such requirements affect the design of the BIG IoT API and the marketplace, as well as any software in the BIG IoT ecosystem. Following this analysis, we have outlined how these requirements will affect the architectural approach of BIG IoT.

Regarding privacy, we have proposed three recommendations that need to be followed by any IoT ecosystem participant: (1) data minimisation, i.e., that a data controller should limit the collection of personal information to what is directly relevant and necessary to accomplish a specified purpose; (2) strong accountability, i.e., to provide mechanisms to securely log any action by any actor dealing with sensitive data; and (3) transparency and easy access, i.e., any data controller should publish transparent and easily accessible data protection policies that clearly show how their data is being processed to the end users. Notice that protecting users' privacy does not necessarily imply added costs. In fact, storing anonymised data can help in saving development and operational costs due to a reduced ASVS level compliance.

Finally, we introduce a use case, and we have analysed it from the perspective of security and privacy. This use case presents an application that helps a user to get to a destination and to easily find nearby parking spots, as well as propose alternative routes by public bus. This use case is being implemented in two pilots of the BIG IoT project, in Barcelona and Berlin/Wolfsburg.

In the future, our research will build up on the recommendations laid out in this article. By implementing various services and applications in the BIG IoT pilots, which all need to follow the security and privacy framework outlined here, we will be able to evaluate our recommendations in terms of feasibility, practicability, and thoroughness. This will lead to sharpened and proven guidelines for the creation of IoT ecosystems in general, which we aim to contribute to our on-going engagement with standardization at W3C's Web of Things group[4].

Beyond the work on security and privacy best practices, we will focus our research agenda towards combining IoT security solutions with Semantic Web [8] technologies. The already available semantic descriptions of services and platforms in the BIG IoT project will enable us to develop ontologies that describe different security aspects. This will allow us to automate the selection of reasonable security measures and options per IoT ecosystem participant.

Acknowledgement. This work is mainly financially supported by the project Bridging the Interoperability Gap (BIG IoT) funded by the European Commission's Horizon 2020 research and innovation program under grant agreement No. 688038. In addition, this work has been partially supported by the MINECO/FEDER funded projects ANFORA TEC2015-68734-R and ARPASAT TEC2015-70197-R and by the Generalitat de Catalunya grant 2014-SGR-1504.

References

1. Open Web Applications Security Project (OWASP). https://www.owasp.org/
2. OWASP Code Review Project second edition guideline. https://www.owasp.org/index.php/Category:OWASP_Code_Review_Project
3. The OWASP Software Assurance Maturity Model (SAMM). https://www.owasp.org/index.php/Category:Software_Assurance_Maturity_Model
4. Worldsensing's Bitcarrier. http://www.bitcarrier.com/
5. Worldsensing's Fastprk. http://www.fastprk.com/
6. Allseen Alliance: Alljoyn framework. https://allseenalliance.org/framework
7. Bartel, M., Boyer, J., Fox, B., LaMacchia, B., Simon, E.: XML Signature syntax and processing, 2nd edn. https://www.w3.org/TR/xmldsig-core/
8. Berners-Lee, T., Hendler, J., Lassila, O., et al.: The semantic web. Sci. Am. **284**(5), 28–37 (2001)
9. Bröring, A., Schmid, S., Schindhelm, C.K., Khelil, A., Kaebisch, S., Kramer, D., Le Phuoc, D., Mitic, J., Anicic, D., Teniente, E.: Enabling IoT ecosystems through platform interoperability. IEEE Software (Software Engineering for the Internet of Things) (2017, forthcoming)
10. EU Legislation: Directive 95/46/ec (1995). https://secure.edps.europa.eu/EDPSWEB/edps/site/mySite/pid/74#data_directive
11. EU Legislation: Directive 45/2001/ec (2001). https://secure.edps.europa.eu/EDPSWEB/edps/site/mySite/pid/86#regulation
12. FTC Staff: Internet of Things: privacy and security in a connected world, January 2015. https://www.ftc.gov/system/files/documents/reports/federal-trade-commission-staff-report-november-2013-workshop-entitled-internet-things-privacy/150127iotrpt.pdf
13. IETF OAuth WG: OAuth 1. https://oauth.net/1/
14. IETF OAuth WG: OAuth 2.0. https://oauth.net/2/
15. Imamura, T., Dillaway, B., Simon, E.: XML encryption syntax and processing. https://www.w3.org/TR/xmlenc-core/

---

[4] http://www.w3.org/WoT/.

16. Jones, M., Bradley, J., Sakimura, N.: JSON Web Signature (JWS). https://datatracker.ietf.org/doc/rfc7515/
17. Jones, M., Hildebrand, J.: JSON Web Encryption (JWE), https://datatracker.ietf.org/doc/rfc7516/
18. Meucci, M., Muller, A.: OWASP testing guideline version 4. https://www.owasp.org/index.php/OWASP_Testing_Project
19. OpenID Foundation: OpenID connect. http://openid.net/connect/
20. Organization for the Advancement of Structured Information Standards (OASIS): Official Wiki of the OASIS security services (SAML) technical committee. https://wiki.oasis-open.org/security/FrontPage
21. OWASP: Application security verification standard 3.0.1. https://www.owasp.org/images/3/33/OWASP_Application_Security_Verification_Standard_3.0.1.pdf
22. OWASP Internet of Things Project: Principles of IoT security. https://www.owasp.org/index.php/Principles_of_IoT_Security
23. Raskin, A.: Privacy icons. https://www.flickr.com/photos/azaraskin/5304502420/sizes/o/
24. Transport Metropolitans de Barcelona: TMB open data. https://www.tmb.cat/en/web/tmb/about-tmb/open-data

Attribute-Based Access Control Scheme in Federated IoT Platforms

Savio Sciancalepore[1,4(✉)], Michał Pilc[2], Svenja Schröder[3], Giuseppe Bianchi[1,5],
Gennaro Boggia[1,4], Marek Pawłowski[2], Giuseppe Piro[1,4],
Marcin Płóciennik[2], and Hannes Weisgrab[3]

[1] CNIT, Consorzio Nazionale Interuniversitario per le Telecomunicazioni,
Parma, Italy
[2] Poznań Supercomputing and Networking Center, IBCh PAS, Poznań, Poland
{michal.pilc,marek.pawlowski,marcin.plociennik}@man.poznan.pl
[3] Cooperative Systems Research Group, University of Vienna, Vienna, Austria
[4] Department of Electrical and Information Engineering (DEI),
Politecnico di Bari, Bari, Italy
{savio.sciancalepore,gennaro.boggia,giuseppe.piro}@poliba.it
[5] Department of Electronic Engineering,
University of Rome Tor Vergata, Rome, Italy
giuseppe.bianchi@uniroma2.it

Abstract. The Internet of Things (IoT) introduced the possibility to connect electronic things from everyday life to the Internet, while making them ubiquitously available. With advanced IoT services, based on a trusted federation among heterogeneous IoT platforms, new security problems (including authentication and authorization) emerge. This contribution aims at describing the main facets of the preliminary security architecture envisaged in the context of the symbIoTe project, recently launched by European Commission under the Horizon 2020 EU program. Our approach features distributed and decoupled mechanisms for authentication and authorization services in complex scenarios embracing heterogeneous and federated IoT platforms, by leveraging Attribute Based Access Control and token-based authorization techniques.

Keywords: Internet of Things · Security mechanisms · Attribute-Based Access Control · Interoperability framework · Macaroons · JSON Web Token

1 Introduction

The term Internet of Things (IoT) was first used by Kevin Ashton in 1999 as a name of a network of RFID devices used to monitor corporate supply chains while simultaneously being connected to the Internet [1]. By 2004 the term had been adopted by most scientific and technological journals like Scientific American [2]. Many proposals of IoT platforms like smart housing, smart stadium or even a smart city have been presented since then [3,4]. In parallel, consortia of

© Springer International Publishing AG 2017
I. Podnar Žarko et al. (Eds.): InterOSS-IoT 2016, LNCS 10218, pp. 123–138, 2017.
DOI: 10.1007/978-3-319-56877-5_8

enterprises and international institutions started to develop protocols and communication standards more suitable for different deployment cases (including machine-to-machine, body area network, industrial telemetric network, and so on). Now, with the incumbent explosion of the IoT in everyday life, heterogeneous application-specific platforms are emerging, often designed as standalone solutions that hardly communicate with each other.

Unfortunately, the fragmentation of the IoT ecosystem resulted in poor cooperation between different IoT platforms in terms of resource sharing and reusability of applications. In the face of the demand for interoperability between different IoT platforms, several international projects like symbIoTe[1], INTER-IoT[2] and bIoTope[3] were launched by European Commission under the Horizon 2020 EU program.

Security is an important cornerstone of all those projects which will impact their success or failure in two aspects: usability and technical implementation. Every solution therefore needs to protect the privacy of users and its resources against unauthorized access and must still provide full functionality.

In distributed (but interoperable) IoT networks, for instance, the protection of resources against unauthorized accesses and the authentication of users requires more sophisticated methods. Conventional computer networks adopt the Role-Based Access Control (RBAC) paradigm. In RBAC a user is assigned a role such as *"administrator"* or *"ordinary user"* that predetermines access rights policies. Unlike RBAC, the Attribute-Based Access Control (ABAC) method of authorization derived from distributed computing relies on the assignment of so-called *"attributes"* to each entity in the system. An attribute may refer either to a user or to a particular resource or to the surrounding environment. An "attribute" is defined as a particular property, role or permission associated to a component in the system. It is assigned after an authentication procedure by the system administrator [5].

In this paper we present a security architecture that enables an ABAC-based controlled access to IoT resources and easily supports a trusted resource sharing among different IoT platforms. The proposed security architecture was developed in the symbIoTe project, which aims at a symbiosis of smart objects across IoT environments. The main contributions of the work are summarized below:

- we describe general requirements for a secure and standardized interoperability framework;
- we identify components to be deployed in the system to manage security issues;
- we provide a baseline architecture for the authentication and authorization among federated IoT platforms;
- we identify different scenarios that require customized security functionalities;
- we design interfaces and interactions among components in the aforementioned architecture;

---

[1] https://www.symbiote-h2020.eu.

[2] http://www.inter-iot-project.eu.

[3] http://biotope.cs.hut.fi.

– we propose two possible technical solutions for the token format, that are Macaroons and JSON Web Tokens (JWTs).

The rest of the paper is organized as in the following.

Section 2 presents security requirements for the IoT interoperability framework and Sect. 3 outlines the state-of-the-art in distributed systems and IoT. Section 4 describes the architecture of the proposed system with focus on security aspects. Following the requirements and architecture, two possible token solutions are presented and compared in Sect. 5. Our efforts, contributions and future work are summarized in Sect. 6.

2   Requirements

In a network of federated IoT platforms it is necessary to provide security of applications, components, platforms and resources at high level. Such complex system must be protected against many security threats: opening doors by an illegitimate user in a Smart Home environment, reading confidential data from remote sensors connected to an industrial sensor network or launching evacuation mechanism in Smart Stadium environment. Due to a huge number of low-power devices, IoT networks are vulnerable to many types of attacks that can cause substantial damages. First of all, every device that is in the radio range can overhear the transmission between sensors, actuators and other devices, thus each message must be protected by cryptographic protocols. Secondly, many IoT devices are not susceptible against malicious updates of their firmware, which can cause damages in a computer network the IoT platform is connected to for instance, letting hackers break into e-mail account[4]. Finally, IoT platforms and networks of IoT platforms are vulnerable to distributed denial-of-service (DDoS) and man-in-the-middle (MITM) attacks. With nearly 20 billion devices that will have been connected to the Internet by 2020 it is crucial to mitigate security threats.

After a detailed analysis of use case sample scenarios, like Smart Home, we defined security requirements of the system. At application layer it is necessary for all devices to adhere to the rules defined within OWASP Internet of Things Project[5]. One of the most important requirement concerning the system is the provision of mutual authentication: this means that not only a user must authenticate with an IoT platform but also an IoT Platform must authenticate with a user. This can help to protect the network against impersonating whole services or servers together with two factor authentication (password and PIN). Authentication and authorization are implemented with cryptographic protocols and primitives. Communication between all system entities must be encrypted to prevent illegitimate access to resources and tampering the data. Another important security mechanism is the validation of input data which is based on

---

[4] http://iotsecurityconnection.com/posts/security-is-a-must-in-everyiot-device.

[5] https://www.owasp.org/index.php/OWASP_Internet_of_Things_Project.

sanitization mechanisms. The countermeasure consists in eliminating potentially harmful characters from user input by means of removal, replacement and encoding the characters. Another important security requirement in such a framework is the implementation of access control through access policies. The users and entities must be able to define constraints that limit the users that are allowed to get access to resources, time constraints and other attribute. The major difficulty here is the capability of granting access to resources from one platform while being a user logged in another IoT platform. To solve this problem, the system must offer identification mechanisms for instance through tokens that store attributes. Another feature that must be implemented in application layer are mechanisms of establishing trust relationships and thus implicitly trust levels, prior to applying security mechanisms for the first time. Information about this must be stored in a secure data store i.e. by Public Key Infrastructure (PKI). Finally, privacy at user level must be preserved. All data about sensors, entities and other resources that are exposed by IoT platform must be anonymized, i.e. devices must not expose details on manufacturer, firmware version and other sensitive data. The details about sensor location must not be exposed unless its owner agrees on it.

3   State of the Art

Guaranteeing user authentication and authorization in distributed computing systems has been always regarded as a concern.

Authorization methods that rely on attributes were widely applied in cloud computing systems, where security policies are supported by the authorization mechanisms of the cloud [6]. More recently a commercial solution of security architecture specifically designed for Supervisory Control and Data Acquisition (SCADA) systems was presented in [7].

A decentralized network of federated IoT platforms like our approach resembles the aforementioned scenarios. The core part of our design is a cloud responsible for seamless connection between sensors, actuators and user applications placed in different IoT platforms. Trust management concerns about the Internet-of-Things were summarized in [8].

A related work showing security threats in IoT was published in 2015 by Sicari et al. [9].

In 2014, the concept of macaroon tokens for decentralized authorization in the cloud was presented [10]. A different authorization method, widely adopted in online purchasing, that is JWT, was described in [11]. Recently, a recommendation for authentication and authorization in IoT was issued in the Authentication and Authorization for Constrained Environments (ACE) IETF Working Group [12]. Scalability and adaptation of access control policies to the environment conditions were proposed as a solution for the Internet of Things [13].

Security concerns were also addressed in EU-funded projects under the 7th Framework Programme (FP7) like OpenIoT[6], SMARTIE[7], RERUM[8], COMPOSE[9] and FI-WARE[10].

To enable a baseline comparison, COMPOSE and OpenIoT are two FP7 projects in which authentication and authorization issues have been tackled by using a centralized solution. In COMPOSE, the middleware is the owner of all registered resources, and manages centrally the authentication of client applications and the issuing of tokens. In OpenIoT, instead, same mechanisms are offered by a Central Authorization and Authentication Service (CAS). This architecture property allows those projects to apply the well-known OAuth 2.0 authorization framework [14]. In our case, instead, the ownership over resources is left to each IoT platform, thus making impossible to apply the OAuth 2.0 paradigm. However, its decoupled logic have been used as a useful starting point for the development of our solution. Finally, while most of the aforementioned projects use ABAC for access control (SMARTIE, RERUM, FI-WARE), COMPOSE uses RBAC and openIoT Lattice-Based Access Control (LBAC). While JSON is being used for issuing tokens in some of the projects, none of them so far considered using macaroons for tokens. However, none of them included attributes directly in the token.

4   Architecture

The reference architecture considered in this contribution is depicted in Fig. 1. It integrates many independent IoT platforms exposing heterogeneous resources. Each IoT platform (thus, each available resource) is registered with a trusted mediator (i.e., the interoperability framework's *core*) which offers advanced mechanisms for enabling platform interoperability and distributed resource access. Moreover, there are applications willing to access the available resources.

To maximize interoperability among platforms our security framework has to deal with different scenarios: applications can be registered to only the trusted mediator, to only one IoT platform, or two (or more) IoT platforms federated with the mediator entity. Therefore, the following target scenarios can be identified:

– **Scenario #1**: an application is registered with an IoT platform and it would like to access resources exposed by the IoT platform where it is registered to. This is the case of a typical cloud system, where applications, services and resources are controlled by the same administrator, without the need to interface with other platforms.

---

[6] http://www.openiot.eu.
[7] http://www.smartie-project.eu.
[8] https://www.ict-rerum.eu.
[9] http://www.compose-project.eu.
[10] https://www.fiware.org.

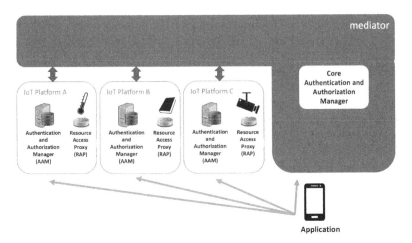

Fig. 1. System architecture.

- **Scenario #2**: an application is registered with the trusted mediator and it would like to access resources exposed by a federated IoT platform. This is the case of a third-party application developer, that implements special applications to access services and resources exposed by a given IoT platform controlled by a different administrator.
- **Scenario #3**: an application is registered with one or more IoT platforms federated with the mediator and it would like to access resources exposed elsewhere in the considered architecture. This is the case of a current sensor in a smart home. To access the data, the system could require an application to register both with Smart-Home and the Service Provider IoT platforms. We refer to this scenario as the *multi-domain access rights composition* paradigm.

4.1 Main Security Rationale

The resource access is handled through the ABAC logic. ABAC is a well-known technique for dealing with access control in distributed environments, which is able to protect sensitive data, applications or services from unauthorized operations by means of efficient, simple and flexible access rules. It is based on *attributes* and *access policy* concepts.

An attribute encodes a specific property, role or permission assigned to an application. Attributes are stored within a digital object, namely a *token*, that certifies the authenticity of both the issuer (i.e., a dedicated component of mediator or IoT platform) and the owner (i.e., the application), additionally to its time validity. Both symmetric or asymmetric cryptography techniques can be used to ensure authenticity and integrity of those tokens. To provide few examples, Macaroons use symmetric keys to generate an Hash-based Message Authentication Code (HMAC) that assures integrity and authenticity of the token, while

JWTs can be created both by using symmetric or asymmetric keys. A thorough description of both solutions will be provided in Sect. 5.

An access policy, instead, enables a fine-grained access control mechanism. In fact, it describes the combination of attributes needed to obtain the access to a given resource. For each resource, a dedicated access policy can be defined. Then, an application in possession of tokens storing a set of attributes matching the aforementioned access policy can successfully obtain the access to the resource. Otherwise, its access request will be denied.

It emerges that the token represents a key element in the resource access mechanism. From the security perspective, it is generated during the authentication procedure and inspected and validated during the authorization procedure. The solution described in this contribution natively offers the decoupling between authentication and authorization processes. This means that authentication and authorization involve different components and are independently executed at different times. An application uses the authentication procedure to authenticate itself within a given domain (like the trusted mediator or an IoT platform federated with the mediator). In case of a successful authentication it obtains a set of tokens storing its own attributes. Then, the collected attributes can be used during the authorization procedure to obtain access to resources. Since an application should not perform the whole authentication process for each resource access, the designed approach allows also for enhanced flexibility and scalability benefits for the whole system.

Note that when an application or component registered in a given IoT platform or in the mediator would like to access resources exposed elsewhere, it could be possible that the attributes that are assigned to it are not valid also in the new domain. Therefore, an *Attributes Mapping Function* is needed to manage the translation between attributes in different platforms. Thanks to the described functionalities, at the same time the interoperability framework works on top and extends the existing security architecture of a given IoT platform, providing procedures for secure communications with foreign IoT platforms and third-party applications.

4.2   Component Description

To practically implement the aforementioned security rationale in each target scenario, the following logical components are introduced:

Platform Authentication and Authorization Manager (AAM): With reference to **Scenario #1** and **Scenario #3**, it handles the authentication procedure for applications registered with the IoT platform. Therefore, it releases *home* tokens storing attributes that describe properties, roles and/or permissions assigned to the application within the platform where it is registered to. It also manages a Token Revocation List (TRL) storing the list of tokens that have been revoked before their expiration. For this reason it may be contacted by any component in the architecture during the *check revocation procedure*.

With reference to **Scenario #2** and **Scenario #3**, the Platform AAM is in charge of (i) verifying the validity and the possible revocation of tokens generated by the AAM component of the mediator or the AAM component of another IoT platform, (ii) performing the attributes mapping functionality and (iii) releasing a new set of tokens, namely *foreign tokens*, usable in the local IoT platform.

Core AAM: With reference to **Scenario #2**, it handles the authentication procedure for applications registered with the mediator. Therefore, it releases *core* tokens storing attributes that describe properties, roles and/or permissions assigned to the application at the mediator side. Moreover, similarly to the Platform AAM, it manages the TRL and can be contacted by any component in the architecture during the *check revocation procedure*.

Resource Access Proxy (RAP): With reference to **Scenario #1**, **Scenario #2** and **Scenario #3**, it holds the URI for obtaining all the resources available in the specific IoT platform, along with the access control policy associated to each of them. Therefore, it receives all the requests for accessing these resources along with the tokens containing the attributes of the requester. Finally, it enforces the access control by checking if the provided attributes satisfies the policy associated with the resource. In the positive case it provides access to the resource itself, otherwise the access is denied.

Note that each AAM component manages authentication and authorization functionalities only for users and applications that are registered within its domain. In this sense, third-party applications that are not registered with any reference IoT platform registers within the core AAM. This design choice provides enhanced flexibility to the interoperability framework, because each AAM does not need to store data related to all possible components in the system. Instead each AAM stores only a limited amount of data. When foreign components tries to access to local services, the reference AAM can interact with the remote AAM to obtain the required information about the application. Finally, we highlight that specific modules devoted to the detection and prevention of system anomalies could be envisaged to work in strict connection with the described modules. However, their design is left for future work.

4.3   Sequence Diagrams

The reference sequence diagrams for the scenarios introduced at the beginning of the section will be provided in what follows. We suppose a previous registration phase, in which each component in the system receives an asymmetric key pair (private/public keys) and a public-key certificate in a trusted way. The public key of the application is also included in the token, to guarantee authenticity of the token itself and to provide the cryptographic material to be used in the challenge-response procedure. Finally, end-to-end security in the communication between described components is guaranteed thanks to the mandatory use of the Transport Layer Security (TLS) protocol [15].

The sequence diagram describing the resource access in **Scenario #1** is depicted in Fig. 2. It is composed by two main steps:

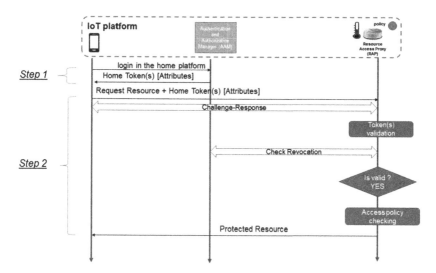

Fig. 2. An application registered in a given IoT platform wants to access resources produced within its home IoT platform.

Step 1: Home authentication. At the beginning, the application performs the login in its home platform by contacting the home AAM and receiving home tokens.

Step 2: Resource access authorization. The application contacts the RAP and delivers the home tokens retrieved in the previous step. The RAP initiates the challenge-response mechanism to verify that the application is the real owner of the tokens. In the case the challenge-response mechanism is successfully completed, the RAP verifies that tokens are valid and that they have not been revoked by contacting its reference platform AAM component. Then it checks the provided attributes against the access policy associated with the requested resource: if the attributes supplied by the applications are enough to satisfy the access policy associated with the resource (according to the ABAC logic) the RAP grants the access to the resource. Otherwise access is denied.

The sequence diagram describing the resource access in **Scenario #2** is depicted in Fig. 3. It is composed by three main steps:

Step 1: Core authentication. At the beginning, the application performs the login with the mediator by contacting the core AAM and receiving core tokens.

Step 2: Foreign authentication. The application forwards core tokens to the AAM component of the foreign platform. The AAM component of the foreign platform initiates the challenge-response mechanism to verify that the application is the real owner of the tokens, thus preventing both replay and impersonation attacks. In the case the challenge-response mechanism is successfully completed, the AAM component of the foreign platform validates the

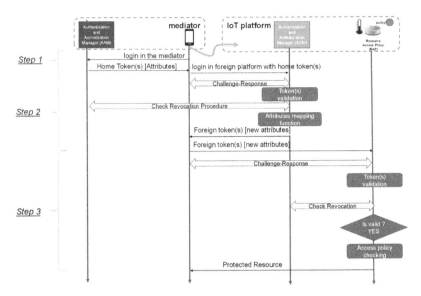

Fig. 3. An application registered in the mediator space wants to access resources in an IoT platform.

tokens, verifies that they have not been revoked by contacting the AAM component of the mediator and performs the attribute mapping function. Then, it generates a new set of foreign tokens and sends them to the application.

Step 3: Resource access authorization. The application contacts the RAP and delivers it the foreign tokens retrieved in the previous step. The RAP initiates the challenge-response mechanism to verify that the application is the real owner of the tokens. In the case the challenge-response mechanism is successfully completed, the RAP verifies that tokens are valid and that they have not been revoked by contacting its reference platform AAM component. Then it checks the provided attributes against the access policy associated to the requested resource: if the attributes supplied by the applications are enough to satisfy the access policy associated to the resource (according to the ABAC logic) the RAP grants the access to the resource. Otherwise the access is denied.

The sequence diagrams describing the resource access in **Scenario #3**, referring to as *multi-domain access rights composition*, are depicted in Fig. 4. Without loss of generality, it is assumed that the application is registered in platforms *IoT_A* and *IoT_B* and would like to gain access to a resource available in the platform *IoT_C*. Also in this case, the procedure has three main steps:

Step 1: Home authentications. At the beginning the application performs the login in the IoT platforms where it is registered to (i.e., *IoT_A* and *IoT_B*). To this end it contacts the AAM component of each platforms for retrieving home tokens.

Step 2: Foreign authentication. The application combines the home tokens and forwards them to the AAM component of the foreign platform IoT_C. The AAM component of the foreign platform initiates the challenge-response mechanism to verify that the application is the real owner of the tokens, thus preventing both replay and impersonation attacks. In the case the challenge-response mechanism is successfully completed, the AAM component of the foreign platform validates the tokens, verifies that they have not been revoked by contacting AAM components of the home platforms (i.e., IoT_A and IoT_B) and performs the attribute mapping function. Then, it generates a new set of foreign tokens and sends them to the application.

Step 3: Resource access authorization. The application contacts the RAP and delivers the foreign tokens retrieved at the previous step. The RAP initiates the challenge-response mechanism to verify that the application is the real owner of the tokens. In the case the challenge-response mechanism is successfully completed, the RAP verifies that the tokens are valid and that they have not been revoked by contacting its reference platform AAM component. Then it checks the provided attributes against the access policy associated with the requested resource: if the attributes supplied by the applications are sufficient to satisfy the access policy associated to the resource (according to the ABAC logic) the RAP grants access to the resource. Otherwise, the access is denied.

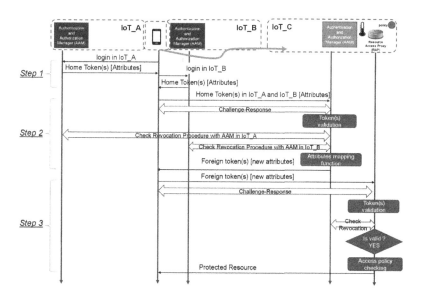

Fig. 4. Multi-domain access rights composition.

5   Planned Implementation

The implementation of the security architecture described in Sect. 4 requires the selection of a suitable token format. Without loss of generality a token is a digital object used as a container for security-related information. It serves for authentication and/or authorization purposes and generally appears as a list of elements. Each element contains an assertion that further specifies properties assigned to the owner of the token. Each token must contain an explicit expiration date, indicating the date until the token can be considered valid. Moreover, the token also contains at the end an element that certifies its authenticity and integrity. Depending on the chosen solution, validating a token could require different procedures.

The following discussion describes two promising technical solutions, candidates for the implementation of tokens in our approach. It does not provide only a comparison between the two approaches, but it illustrates also, specifically in Sect. 5.3, how they can be modified in order to fit within the proposed architecture.

During implementation of the framework itself one or a combination of the following technologies will be used. We aim at a flexible solution where platforms, applications and other use cases can decide which of the following technologies they want to use.

5.1   Macaroons

Macaroons are a new kind of authorization credential developed by Google [10]. As bearer credentials, they serve a similar purpose as cookies in World Wide Web, but they are more flexible and provide better security. Macaroons are based on a construction that uses nested, chained MACs (e.g., HMAC) in a manner that is highly efficient, easy to deploy and widely applicable. They allow authority delegation between bearers with attenuation and contextual confinement. Each field embedded within macaroons structure, i.e. the caveat, restricts both the macaroons' authority and the context in which it may be used (e.g. by limiting the permitted actions and requiring the bearer to connect from a certain IP address and to present additional evidence such as a third-party signature). Macaroon caveats are plain-text readable. Macaroons also contain a list of *AND* conditions. Its bearer is authorized to perform an operation *AS LONG AS condition1, condition2, ..., conditionN* hold true. The main (root) macaroon which allows for everything gets successively attenuated with those conditions. Each of them is signed with an HMAC function.

By considering the reference architecture shown in Fig. 1, three possible tokens can be introduced: *root macaroons*, *platform macaroons*, and *application macaroons*. The mediator creates a root macaroon by calculating the HMAC function of a random nonce and its secret key. Note that the output of the HMAC function must be shared among AAMs of federated platforms for verifying the authenticity of tokens received during the resource access procedure. Starting from the root macaroon the mediator also generates platforms macaroons. The platform macaroons are signed by the HMAC function with the key being the previously calculated HMAC value. Then, each platform can autonomously generate application macaroons by following the same process.

An application, after obtaining the macaroon from the AAM component of its platform, may further attenuate it and therefore authorize others to perform actions in its name. For example a Smart Home System mints its owner a full access token. The owner can go on vacation and want the neighbor next door to be able to operate the windows and doors but not the garage. The owner can do so by attenuating his/her own token and handing it to the neighbor.

5.2   JSON Web Tokens

JWT is an open industry standard widely used in today's Internet to deal with authentication and authorization issues [11]. It contains a set of *claims*. A claim is a specific certified statements related both to the token itself or to the entity that is using it. Typically, these claims are encoded in the JavaScript Object Notation (JSON) format, thus easily allowing system interoperability. A claim is identified with a specific name: it is possible to distinguish between *Registered Claim Names*, that are names defined and standardized in the reference document and *Private Claim Names*, that represent extensions that a developer could choose for his/her own system.

The cryptographic force of the JWT resides in the sign field, stored at the end of the token. It can be generated through symmetric or asymmetric cryptography techniques and allows to verify the authenticity of the token, i.e. generation by a trusted entity, as well as integrity, in the sense that no one could modify its content without invalidating it.

Each JWT contains a header that provides information about the type of the token and the algorithm used to build the sign of the token. It contains also a body, encoding a set of claims for this token, and finally - a sign containing the cryptographic validation of the token and generated as stated in the header.

Registered Claim Names carried in the body of the JWT include *iss*, that uniquely identifies the entity that issued the token, *sub*, which uniquely identifies the entity for which this token has been released (it is a key field when a token needs to be used also for authentication purposes), *exp*, indicating the expiration time, after which this token should not be used and processed by any entity in the system; *nbf*, that identifies the time in which this token becomes effectively valid and can be processed by any entity in the system, *iat*, identifying uniquely the time in which this token has been created and, finally, *jti*, that is the unique identifier of the token.

JWT fits perfectly within the reference architecture described in Sect. 4. From a cryptographic perspective, the only requirement for its adoption is the deployment of a public-key infrastructure, which issues a private/public key pair to each entity in the system.

AAMs uses their private keys to generate and sign tokens. Any entity in the system that receives a token could easily verify its authenticity by gathering the public key of the issuer of the token (specified in the token itself).

Important features such as the support for an expiration date are integrated by JWT thanks to the definition of the *exp* claim.

Also, each token can be easily associated to a given entity in the system through the *sub* claim. More in detail, the public key of the owner of the token can be embedded in this claim. This can be used in the challenge-response procedure described in Sect. 4 to prove the possession of the respective private key and verify that the application using the token is effectively the entity for which the token has been generated. This procedure avoids replay attacks.

Finally, the JWT can be easily extended to support the carrying of attributes associated to the ABAC logic, thanks to the possibility to integrate customized Private Claims.

5.3   Usage of Tokens in Our Proposal

To conclude, the comparison between the user macaroon and the user JWT, as they can be modified to be included in the described architecture, is reported in Fig. 5.

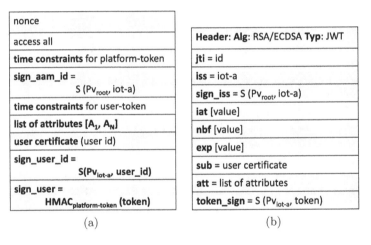

Fig. 5. Details of the content of (a) Macaroons and (b) JWT tokens for the designed architecture.

The application macaroon is shown in Fig. 5(a). It has a hierarchical structure. The nonce is used as a unique identifier of the token. The second, third and fourth caveats are related to the AAM of the platform in which the application is registered to. In particular, the ID of the AAM is signed through the private key of the mediator entity (PV_{root}). The remaining lines are dedicated to the application. They contain the list of attributes assigned to the application, its public key certificate and, finally, the sign on the user ID by the AAM that issued the token, through its private key (PV_{iot-a}). The last line is needed to assure that the AAM is the unique component able to sign the token. Finally, the last caveat is the chained HMAC of all the caveats in the token, performed starting from the output of the HMAC function of the root macaroon.

The application JWT is shown in Fig. 5(b). Also in this case we can identify the part related to the issuer of the token, certified through the sign with the private key (PV_{root}), and the part related to the owner of the token. The private claim *att* is introduced to encode the information about the list of attributes possessed by the application. Finally, the sign of the whole token is performed through the private key of the issuer (PV_{iot-a}) without the need of a symmetric shared secret.

Note that both token types have a limited time validity. After the expiration of that date, the token must be renewed through a new authentication procedure.

The evaluation of pros and cons of macaroons tokens and JWTs as well as their suitability for the proposed scenario is left for future work.

6   Conclusions and Future Activities

In this paper we presented the baseline security architecture for an interoperability framework among IoT platforms, developed within the H2020 EU project symbIoTe. First, we described our general system requirements derived from use cases and requirements. We illustrated our approach for a standardized IoT architecture with the focus on security. Current plans for implementation foresee the usage of the ABAC paradigm, through Macaroons or JWT tokens. However, this paper so far provides a basic architecture and a proposal for some technical solutions to realize a security framework. Future work will aim first at implementing the proposed security architecture within the H2020 symbIoTe project, and conducting a deep performance evaluation of the two approaches described for tokens. Also, a solution for detecting anomalies in the system will be developed and device-level security will be carefully considered. Our main action point for future work, however, will be on the implementation of the aforementioned secure interoperability framework (and thus validation of the architecture concept).

Acknowledgments. This work is supported by the H2020 symbIoTe project, which has received funding from the European Union's Horizon 2020 research and innovation programme under grant agreement No. 688156. The authors would like to cordially thank the entire symbIoTe consortium for their valuable comments and discussions.

References

1. Ashton, K.: That Internet of Things thing. RFID J. **22**, 97–114 (2009)
2. Gershenfeld, N., Krikorian, R., Cohen, D.: The Internet-of-Things. Technical report, Scientific American (2004)
3. Gross, M.: Smart house and home automation technologies. Technical report, Encyclopedia of Housing (1998)

4. Mohanty, S.P., Choppali, U., Kougianos, E.: Everything you wanted to know about smart cities. IEEE Consum. Electron. Mag. **5**(3), 60–70 (2016)
5. Hu, V., Ferraiolo, D., Kuhn, R., Schnitzer, A., Sandlin, K., Miller, R., Scarfone, K.: Guide to Attribute Based Access Control (ABAC) definition and considerations. NIST special publication 800-162. NIST, January 2014
6. Khan, A.: Access control in cloud computing environment. ARPN J. Eng. Appl. Sci. **7**(5), 613–615 (2012)
7. Juniper-Networks: Architecture for secure SCADA and distributed control system networks. Juniper Networks White Paper (2010)
8. Yan, Z., Zhang, P., Vasilakos, A.: A survey on trust management for Internet of Things. J. Netw. Comput. Appl. **42**, 120–134 (2014)
9. Sicari, S., Rizzardi, A., Grieco, L., Coen-Porisini, A.: Security, privacy and trust in Internet of Things: the road ahead. Comput. Netw. **76**, 146–164 (2015)
10. Birgisson, A., Gibbs Politz, J., Erlingsson, U., Lentczner, M.: Macaroons: cookies with contextual caveats for decentralized authorization in the cloud. In: Proceedings of the Conference on Network and Distributed System Security Symposium (2014)
11. Jones, M., Bradley, J., Sakimura, N.: JSON Web Token (JWT). RFC 5719, IETF, May 2015
12. Seitz, L., Selander, G., Wahlstroem, E., Erdtman, S., Tschofenig, H.: Authorization for the Internet of Things for constrained environments draft-ietf-ace-oauth-authz-04. Internet draft, IETF (2016)
13. Hennebert, C., et al.: IoT governance. privacy and security issues. Technical report, European Research Cluster on the Internet of Things, January 2015
14. Hardt, D.: The OAuth 2.0 authorization framework. RFC 6749, IETF, October 2012
15. Dierks, T., Rescorla, E.: The transport layer security protocol Version 1.1. IETF, April 2006

Platform Performance and Applications

Data Ingestion and Storage Performance of IoT Platforms: Study of OpenIoT

Alexey Medvedev[1](✉), Alireza Hassani[1], Arkady Zaslavsky[2,3],
Prem Prakash Jayaraman[4], Maria Indrawan-Santiago[1], Pari Delir Haghighi[1],
and Sea Ling[1]

[1] Faculty of Information Technology, Monash University, Melbourne, Australia
{alexey.medvedev,ali.hassani,maria.indrawan,pari.delir.haghighi,
chris.ling}@monash.edu
[2] CSIRO, Data61, Melbourne, Australia
arkady.zaslavsky@csiro.au, arkady.zaslavsky@acm.org
[3] ITMO University, St. Petersburg, Russia
[4] Swinburne University of Technology, Melbourne, Australia
pjayaraman@swin.edu.au

Abstract. Internet of Things is a very active research area with great commercialisation potential. The number of IoT platforms is already exceeding 300 and still growing. However, performance evaluation and benchmarking of IoT platforms are still in their infancy. As a step towards developing a performance benchmarking approach for IoT platforms, this paper analyses and compares a number of popular IoT platforms from data ingestion and storage capability perspectives. In order to test the proposed approach, we use the widely used open source IoT platform, OpenIoT. The results of the experiments and the lessons learnt are presented and discussed. While having a great research promise and pioneering contribution to semantic interoperability of IoT silos, the experimental results indicate OpenIoT platform needs more development effort to be ready for any substantial deployment in commercial IoT applications.

Keywords: Internet of Things (IoT) · Platform · Data management · Storage · Ingestion · Evaluation · Benchmarking

1 Introduction

According to the Gartner's report [1] the market of Internet of Things (IoT) platforms has been continuously emerging. Gartner defines an IoT platform as "a software suite or a Platform as a Service (PaaS) cloud offering that monitors, and may manage and control, various types of endpoints, often via applications end users build on the platform. It facilitates operations involving IoT endpoints and integration with enterprise resources." The capabilities of a typical IoT platform include device management, data processing, storage and management, event processing, instruments for application

I. Podnar Žarko et al. (Eds.): InterOSS-IoT 2016, LNCS 10218, pp. 141–157, 2017.
DOI: 10.1007/978-3-319-56877-5_9

development, security, device interoperability and interfaces for administrators, developers and users [1]. Existing platforms seriously vary in their prices, functionality, limits and performance.

Performance evaluation of an IoT Platform [2] is a challenging task due to the variety of systems and their components. The discussion of developing a benchmarking system for IoT platforms has started recently, but the process is still in its infancy. However, developers and companies have an urgent need to compare solutions in order to understand development trends and making an appropriate choice while selecting an IoT platform. Data management infrastructure is one of the core components of an IoT platform and it has a significant influence on the overall performance. This makes the problem of benchmarking the process of data ingestion, storage and retrieval important.

In this paper, we analyse what architectural components are used across different IoT platforms and how they influence the overall performance. In particular, we are interested in the ingestion and storage components of IoT platforms and propose a set of IoT platform benchmarking metrics. Our analysis show, the performance of an IoT system's storage components depend not only on the underling storage technology, but also on the ingestion pipeline, messaging queue and many other components of the platform. Additionally, decisions about how to store, cache and access different types of data have significant influence on the overall performance. We use the proposed benchmarking metrics to run a series of experiments to evaluate and test the ingestion and storage performance of the widely used open source platform - OpenIoT. Finally, we provide a detailed analysis of the experimental outcomes discussing OpenIoT's data ingestion and storage performance.

The paper is organised in the following way: in Sect. 2, we provide information about approaches to data storage taken by different IoT platforms. Section 3 describes how the OpenIoT platform manages ingestion, storage and retrieval of data. This description if followed by a series of tests, which identify limits and bottlenecks in current implementation of OpenIoT. Section 4 contains discussion on IoT platforms benchmarking and Sect. 5 concludes the paper and defines directions for future work.

2   Overview of Data Management in Existing IoT Platforms

In this section, we analyse how different IoT platforms deal with data management (storage) issues. We have grouped the platforms into two categories: commercial IoT platforms and academy/research projects. This division is caused by seriously different approaches applied by these two groups. The difference in targets and approaches makes it hard to compare projects from different groups, but we believe that best practices from one group can be applied in another for improving the functionality and performance.

Avoiding technologies that were not seriously tested in real-world applications, focusing on security, performance and cost efficiency are essential features of all commercial solutions. These solutions are often cloud-based and use a Software as a Service (SaaS) or PaaS model. Their main target is to allow customers straightforward and rapid development of applications that will connect companies' "things" together. These solutions provide interoperability regarding various sensors and protocols, but the problem of horizontal data exchange between companies, state organizations, and individuals is not addressed. Research projects, on the other hand, provide open-source software that can be deployed on-premises and maintained by organizations themselves. These projects often focus on semantic and horizontal interoperability and while providing cutting edge functionality in some aspects can lack functionally or performance in others.

In this paper, we are focusing on data management approaches, for instance, how data is ingested, stored, indexed, cached and retrieved by the IoT platform.

2.1 Commercial IoT Platforms

Predix [3] is a PaaS IoT platform developed by General Electric and aiming to provide services for data collection and processing in the area of Industrial Internet of Things. Predix has a catalogue of provided services that includes a set of tools for data management. Predix is not trying to produce a one-fits-all solution and propose to use one or several services best suited for the current task. These services are (i) Asset Data, (ii) Time Series, (iii) SQL database, (iv) Blobstore, and (v) Key-Value store. RabbitMQ – a message queue based on AMQP protocol, organizes communication of components. Asset data is a set of models used to describe machines and instances, which are created basing on these models. Time Series service provides means for efficient ingestion, distribution and storage of sensory data, including indexing for making fast queries. Predix uses a graph database for its asset service to store data as RDF triples. A special Graph Expression Language (GEL) is used for data retrieval [4]. SQL database service is built on top of well-known open-source PostgreSQL database. Blobstore service provides means for storing and retrieving any amounts of binary data and ensures high availability and horizontal scalability. Key-Value store service is based on open-source Redis project and serves as an advanced cache store. Predix uses a hybrid (or polyglot persistence) storage solution, but all the responsibilities on choosing the right options are left to the application developers.

Data services is a promising feature of the platform that is in the beta stage. There are only two services available: Places data services and Seismic data services. This is a remarkable step to horizontal IoT solutions. The platform provides easily accessible data from external data sources or sensors to application developers making it possible to adapt industrial automation solutions to detected earthquakes or other accidents.

Tibbo AggreGate [5] is an example of a commercial non-cloud IoT platform. All data is logically separated into two groups: (i) configuration and (ii) events. This approach helps in providing flexibility of data storage in case of adding new business objects. Configuration data can be stored in almost any enterprise-grade relational database that supports JDBC connectivity, key-value database or in a file-based storage. In case of a relational database the AggreGate platform includes an embedded database or a preconfigured version of MySQL. AggreGate provides means for database clustering for achieving high availability. Key-value integrated storage is recommended for scenarios, which need clustering together with a high update rate. File-based storage can be used in environments with limited resources. Event data can be stored in a relational database, NoSQL database or in-memory storage. RDBMS puts some limits on insertion performance. NoSQL database provides horizontal scalability, high insertion rates and failover facilities. Memory storage is proposed for use in some of the embedded installations. Approximate estimated insertion rates for a relational database are about 500–2000 events per second and 10–20 thousand events per second for a NoSQL database on a standard server node [6]. AggreGate also provides functionality for building a failover cluster and achieving high availability.

ThingWorx [7] is a cloud-based platform enabling developers to build solutions for IoT. It provides three main ways for data storage: (i) Data Tables, (ii) Streams and (iii) Value streams. ThingWorx also uses concepts of an InfoTable and a DataShape.

The InfoTable is a JSON document in which all the objects share the same properties. InfoTables are fast in-memory objects and are recommended for storing temporary data. The DataShape specifies what property names are required in an object and what types they have. This means a DataShape represents a schema for defining a "thing".

The concept of a DataTable in ThingWorx is similar to a table in relational databases, but columns are defined by a DataShape. A DataTable supports the creation of indexes upon its properties. It is recommended to build an index for each common request for achieving high performance and to use DataTables when it is expected to have not more than 100000 rows in it. Storage of time series data is facilitated by streams. A stream consists of a timestamp and additional properties defined by a DataShape. For dealing with things-driven models, it is recommended to use Value Streams, which have some differences with ordinary Streams. Value Streams provide persistence for associated property and return only property values on request. On the contrary, a stream returns a whole row when querying a single column.

Amazon AWS IoT is a cloud based platform that makes use of all the impressive technological stack provided by Amazon. Communication between devices and cloud is organized by a Device Gateway which supports the publish/subscribe approach. The Rule Engine provides means for configuring rules for filtering and transforming incoming events. This configuration includes routing of data to various supported databases, messaging queues, AWS Lambda and other services. Registration and monitoring of connected devices can be made via Device Registry. Configuration of processing rules for the device is performed in a JSON document consisting of an SQL statement and an action list.

Amazon's IoT solution uses storage technologies provided by Amazon Storage Services. Full description of its capabilities is not possible in this paper due to space limitations. Amazon Storage Services focus on providing scalability, availability and elasticity for mostly well-known storage technologies and promote a so-called NoDBA approach, which reduces operational costs for customers. The variety of provided storage services includes Amazon DynamoDB, Amazon RDS, ElastiCache and ElasticSearch. Amazon DynamoDB is a cloud managed NoSQL key-value store, but a version for on-premises installations is also available. Amazon Relational Database Service (Amazon RDS) can use any of the six most popular relational databases. Amazon's in-memory data store cloud service is represented by so-called "ElastiCache". This service can significantly improve system performance by reducing the number of slow disk reads. ElastiCache is based on two popular open-source in-memory engines: Redis as an in-memory data store and Memcached as a system for object caching. For such use cases as device-log analysis and real-time monitoring of applications Amazon recommends the ElasticSearch service that is based on a famous cognominal search engine. Amazon IoT platform uses a messaging system based on Kafka-based named "AWS Kinesis" for event-broadcasting. Capturing and loading streaming data is performed by Kinesis Firehose and analytical processing of streaming data is done by Kinesis Analytics [8, 9]. Amazon AWS IoT introduces the "thing shadow" or "device shadow" concept. A special "Thing Shadows" service is responsible for managing fast and easy access to a JSON document with a current state of device that was last reported to the platform.

IBM Watson IoT platform relies on the IBM Cloudant database. It is a cloud fully managed document-oriented database sharing many common features with Apache CouchDB. IBM recognizes the need for flexible storage solutions, but by now their solution is mostly document-oriented. Describing plans for future, IBM's specialists state that variety of tasks causes different requirements to latency, scalability, cost and performance, raising the necessity for different storage solutions [10]. By now, data from devices can be stored in two formats. If the API receives a valid JSON, it is stored in the same way. In another case, the data is saved as a base64 encoded string inside the payload field of a JSON document. Recently IBM introduced a feature named "Last Value Cache". As the most common request to an IoT device is about its current state, it makes sense to provide a way for answering such requests in the fastest possible way. These cached values are stored for 12 months and can be retrieved using the standard API.

2.2   Other Platforms

Many other PaaS platforms including Carriots, Xively, Zatar, Realtime.io and Flow-things.io provide only APIs for uploading or downloading data and do not provide information about their backend and storage architecture.

2.3   Academic/Research Projects

The **FIWARE** community is aiming to create an open ecosystem that will enable development of Smart Applications. This ecosystem is based on royalty-free standards and covers a wide range of tasks. Software for the category of tasks is grouped into a module called "generic enabler" (GE) [11].

FIWARE provides several generic enablers for dealing with various types of storage. The central module of FIWARE ecosystem is the Orion Context Broker. This component uses a connector called "Cygnus" that is responsible for persisting or retrieving data from a specific storage. The current release of Cygnus can communicate with HDFS, MySQL, PostgreSQL, CKAN, MongoDB, Comet, Kafka, DynamoDB and CartoDB.

Time series data in FIWARE ecosystem is managed by a component called Comet or Short Term History (STH). This component deals with the storage, retrieval, and removal of raw time series data as well as aggregated context information. This component relies on MongoDB as the datastore.

Semantic Application Support (SAS) GE provides the possibility for developing applications based on Semantic-web technological stack. In [12] Ramparany et al. suggest that OWL and other Semantic technologies can help in solving such problems as (i) Semantic data interoperability, (ii) data integration and abstraction, (iii) data discovery, and (iv) reasoning. FIWARE developers admit that despite large investments and development of markup and query languages the progress with penetration of Semantic web technologies into the market is still too slow. They identify several reasons, which include both technical and commercial problems. SAS GE tries to solve technical and engineering problems, namely (i) scalability, (ii) performance, (iii) distribution (iv) security, (v) lack of methodologies and best practices, and (iv) lack of development instruments. The GE consists of a GUI client and server-side components that are responsible for storing and managing ontologies. Server-side components provide scalable and secure ways to publish and retrieve metadata as well as instruments for managing the infrastructure and data.

The data layer of SAS GE consists of relational database that stores information about ontology documents, and a Knowledge Base that supports OWL-2RL. By now, there is no knowledge base-independent solution and the knowledge base is implemented as a combination of Sesame and OWLIM [13].

OpenIoT is an open source IoT platform, which includes a set of novel functionalities [14], namely:

- Incorporation of IoT data and applications inside cloud computing infrastructures;
- Providing a secure access to semantically interoperable applications;
- Enabling and supporting discovery of sensors and data at run-time;
- Supporting mobile sensors and corresponding QoS parameters.

In OpenIoT the registration, data acquisition and deployment of sensors are managed by X-GSN. X-GSN is an extension of the GSN [15], which is responsible for semantically annotating both sensor data and metadata. Virtual sensor is the main fundamental concept in X-GSN, which is capable of representing any abstract entity (e.g. physical devices) that collects any parameters. In order to make a virtual sensor accessible from the rest of the OpenIoT platform, each virtual sensor needs to register within the Linked Sensor Middleware (LSM). LSM is another core component in OpenIoT that is responsible for handling the sensor data delivery chain. In this regard, LSM transforms and annotates (based on the supported ontologies) the data coming from virtual sensors (through X-GSN) into a Linked Data representation i.e. RDF, and stores it in the database. The OpenIoT platform relies on OpenLink Virtuoso (it is also known as Virtuoso Universal Server) as the main database. OpenLink Virtuoso is a hybrid database engine that combines the functionality of a traditional RDBMS, ORDBMS, virtual database, RDF, XML, free-text, web application server and file server functionality in a single system [16]. According to information on the website, Virtuoso can handle the insert rate of 36 K triples per second on a single 4-core machine. The architecture of OpenIoT data platform is illustrated in Fig. 1.

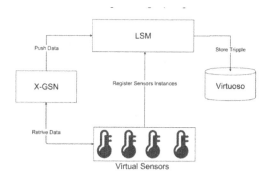

Fig. 1. OpenIoT data platform architecture

2.4 Discussion

After analysing the data storage capabilities of several commercial and research IoT platforms, we have identified that such platforms mostly tend not to limit developers in their choice of a data storage format. Some IoT platforms introduce their own storage technologies; others offer well-known open source or commercial solutions. Mostly these platforms offer the following storage types: (i) in-memory, (ii) document-oriented, (iii) column-oriented, (iv) relational, and (v) RDF. Organization of blob storage is performed using OpenStack Swift. Scalability and high performance of message queueing are achieved by using technologies like Apache Kafka, RabbitMQ, or ZeroMQ. For Big Data processing IoT platforms usually rely on Apache Hadoop or Apache Spark. Research prototypes often use RDF or OWL, but this trend is still mostly avoided by commercial companies due to the issues with scalability and low industry penetration of Semantic Web technologies.

It is also worth noticing that some of the discussed platforms are developing and introducing new features at a very high pace so that we can expect serious changes in the market in the nearest future. Table 1 displays key features of several popular IoT platforms from an IoT data storage angle. It is easy to notice that several platforms rely on multiple technologies for data ingestion and storage instead of using one approach for all types of data.

Table 1. Underlying storage technologies of IoT platforms

Solution/Feature	IBM IoT	Predix	ThingWorx	Tibbo AggreGate	Amazon IoT	FI-Ware	OpenIoT
C/L	C	C	C	L	C	H/L	L
TS	IBM Cloudant	Column storage	Streams (Cassandra)	NoSQL	DynamoDB	Comet	-
R	-	PG	PG (any JDBC)	Embedded, MySQL or any other enterprise-grade DB	RDS (6 databases)	MySQL, PG	Virtuoso
D	Cloudant (Couch DB)		JSON	Integrated NoSQL	ElasticSearch	MongoDB	-
KV	-	Redis	-	Integrated	DynamoDB	Comet	-
IM	-	Redis	InfoTables	Integrated	Redis/Memcached	-	-
MQ	MQTT	RabbitMQ	MQTT	-	Kinesis	RabbitMQ	-
HS	+	+	+	+	+	+	-
SW	-	RDF Graph	-	-	-	SAS GE, Sesame and OWLIM, Apache Jena, SDB	Virtuoso (RDF graph)

C = Cloud, L = Local/on-premises deployment, R = Relational, D = document-oriented, KV = Key Value, IM = in-memory, BD = big data, HS = Horizontal Scalability, SW = Semantic Web stack, MQ = Messaging Queue, PG = PostgreSQL, "-" = Not described in documentation or not implemented

Based on the study of data management techniques and tools presented earlier, in Sect. 3, we present the benchmarking metrics used to evaluate the performance of IoT platforms from the perspective of storage and data ingestion. We use the OpenIoT platform in order to conduct benchmarking studies using the identified metrics.

3   Benchmarking Metrics and Evaluation of OpenIoT

In this section, we discuss the evaluation of data ingestion and storage capability in the OpenIoT platform. In particular, we propose a set of benchmarking metrics that are then used in the experimental evaluation of the OpenIoT platform. We refer to the term "ingestion" to describe the whole process of loading data to the system through a set of components. We use the term "injection" for describing the generation of an individual sensor measurement.

3.1   Benchmarking Metrics

Two different metrics are proposed in order to benchmark the performance of the IoT platform. These metrics are Ingest per Second (IPS), and Resource usage.

- *Ingest per second* is defined as the total number of ingested data divided by the injection duration. For each experiment, we calculate the number of total ingested data points and divide it by injection duration, which is 60 s. This metric determines the performance of the OpenIoT platforms' data ingestion capability.
- *Resource usage* shows the amount of resources (CPU usage, and Memory) consumed by the IoT platform during the execution of each experiment. Resource usage metrics are collected every five seconds when the data injection starts until it ends. As we mentioned in Sect. 2.3, OpenIoT data ingestion pipeline consists of three main components, X-GSN, LSM, and Virtuoso. Therefore, in this paper, we are only interested in the resource usage of these three components. However, since LSM runs on JBoss, we logged and represented the JBoss resource usage instead of LSM.

3.2 Experiments Settings

To evaluate the performance of data ingestion, a large set of sensors with different injection rates is needed. However, due to limitations in accessing resources (sensors) we used synthesized sensor data streams to perform the large-scale evaluation.

In OpenIoT, the process of generating virtual sensors consists of two main steps, defining sensor types, which will be stored as an extension to the OpenIoT ontology in the LSM server, and creating new instances of the sensor types. For the purpose of this study, we are only interested in the performance evaluation of data ingestion. Therefore, increasing the number of sensor types does not affect the results of our experiments as the type and exact value of the ingested data points does not have a considerable impact on IPS. Accordingly, we predefined only one sensor type and used it for all the experiments. This sensor type is a weather station that observes temperature and humidity. For creating a virtual sensor, two files with the same name (sensor name) are needed to be generated inside the X-GSN virtual-sensor folder. These files are Virtual Sensor Description file (VSD), and Virtual Sensor Metadata file (VSM). The VSM is used for associating metadata with a virtual sensor. On the other hand, VSD is an XML file that contains the selection and the parametrization of the virtual sensor and the corresponding wrapper, which is responsible for data acquisition. Based on the requirements of our experiments, we implemented a new time driven wrapper called Simulation-Wrapper. The Simulation-Wrapper first retrieves the injection rate, injection duration, and sensor output format from the VSD file. Then it periodically (interval is based on the injection rate) randomly generates the sensor's data output and sends it to the LSM server for the injection. In all conducted experiments, the injection duration is set to one minute. We implemented a sensor simulator for generating and registering virtual sensors.

To evaluate the performance of data injection in OpenIoT platform, 121 experiments have been conducted. In these experiments, we have studied the impact of increasing the data ingestion rate by (a) varying the number of sensors while the injection rate is fixed; and (b) varying the injection rate while the number of sensors is fixed. First, we have fixed the number of sensors to 1 and increase the injection rate from 1 to 100 (with the increment step set to 10). Then we repeat the experiments ten more times by increasing the number of sensors by 5 in each iteration. This procedure is shown in the following pseudocode:

```
Sensor_Number = 1
For Sensor_Number <= 50
  Injection_Rate = 1
  For Injection_Rate <= 100
    Run-Experiment(Sensor_Number, Injection_Rate)
    Injection_Rate = Injection_Rate +10
  End For
  Sensor_Number = Sensor_Number + 5;
End For
```

Furthermore, to get more precise results and eliminate unwanted factors that could affect the experiments results, we restart the Virtuoso, JBoss, and X-GSN servers before starting each experiment. The detailed steps of the designed experiments are depicted in Fig. 2.

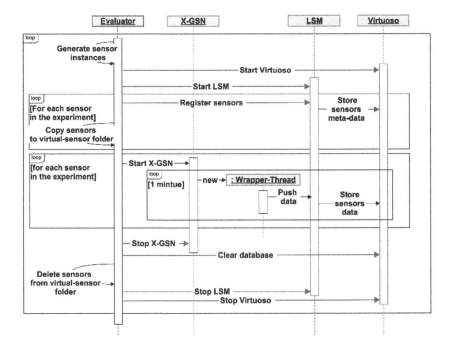

Fig. 2. Evaluation sequence diagram

3.3   Experimentation Environment

The OpenIoT platform is deployed on the JBoss application server hosted on a VirtualBox. The experiments were carried out in the following environment: MacBook Pro (Mid 2015), Oracle VirtualBox 5.1.4, Intel i7-4770HQ CPU @ 2.20 GHz × 4 cores, 8 GB 1600 MHz DDR3 memory, 40 GB Flash Storage, Ubuntu 16.04 (64-bit) OS.

3.4   Experimentation Results

In this section, we report the results of each experiment for the synthetic dataset regarding ingest per second (IPS), and Resource usage

Ingest per Second. Figure 3 shows the IPS versus injection rate where the number of sensors is fixed to 1, 5, 10, 15, and 20. This graph indicates that when the injection rate is less than 30, the increment of injection rate has a direct impact on IPS. In contrast, a sharp drop can be observed when the injection rate is between 30 and 40 injections per second. Moreover, when the injection rate increases to more than 50, the IPS almost remains constant. Overall, the data indicates that increasing the injection rate improves the IPS until the platform reaches its maximum capacity, where the injection rate is equal to 30, and from that point, IPS immediately drops (due to the extra overhead) and then remains unchanged with further increase of injection rate.

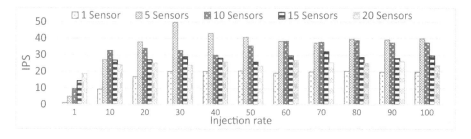

Fig. 3.   Data ingestion performance with different injection rates

Similar data is represented in Fig. 4. This figure shows the IPS versus number of sensors where the injection interval is fixed to 1, 10, 20, 30, and 40 injections per second. While we expect to observe an increase in IPS as a result of raising the sensors number, at some points, potentially when the sensors number goes up, we observe a decrease in IPS. We can see that when the number of sensors is 5, the IPS reaches its maximum (49.6 ingestion per second). This is followed by a gradual decline in IPS when the number of sensors is between 5 and 20. Then the IPS almost remain unchanged for the rest of the trial.

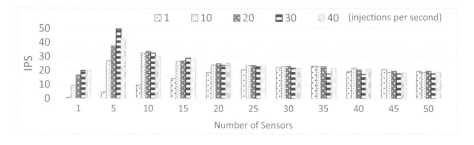

Fig. 4.   Data ingestion performance with varying number of sensors

Based on the represented data, we can conclude that OpenIoT does not perform efficiently when the total injection rate increases to more than 50 IPS.

The last chart we want to discuss in this section is represented in Fig. 5. This chart shows how the IPS ratio changed among all the conducted experiments and has two graphs; the blue graph (single line) shows the actual IPS and the red graph (double line) represents the expected IPS rate. Because the IPS rate is positively skewed, we used a logarithmic transformation (base 10) to normalize the data. The labels on the horizontal axis represent the injection ratio and between each two horizontal gridlines, the number of sensors increases from 1 to 50 (with increment step set to 5). To estimate the expected IPS for each experiment, the following equation is used:

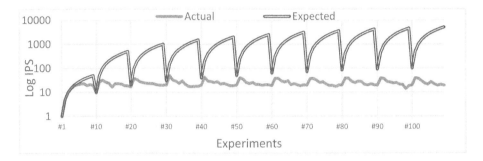

Fig. 5. Data ingestion performance for different experiment settings described in Sect. 3.2 (Color figure online)

$$Expected\ IPS_{1injection\ rate=m}^{\#Sensors=n} = m * n$$

As the double line graph shows, we expect IPS to increase rapidly in each section (between horizontal gridlines) while the number of sensors increasing. Furthermore, we expect an overall increase in IPS since the injection ratio grows. However, the actual value appear to be different from expected. In the first section, when the injection rate is 1, the actual value is almost same as the expected value in very first points. However, when the number of sensors get larger, the gap between actual and expected value starts growing. In the case of second section (inject per second = 10), IPS first increases (when number of sensors is between 1 and 5) and then start decreasing. This pattern repeats in the rest of the trial and the overall IPS ratio remains steady while the distance between the actual and expected value increase dramatically.

Resource Usage. Figure 6 shows the memory usage among all the conducted experiments and has three graphs: the blue graph (dotted) shows the Virtuoso memory usage, the red graph (single line) represents the X-GSN memory usage, and the green graph (double line) indicates the JBoss memory usage. The labels on the horizontal axis represent the injection ratio and between each two horizontal gridlines, the number of sensors increases from 1 to 50 (with increment step set to 5). As it can be observed, both X-GSN and JBoss memory usage fluctuate without any specific pattern. Therefore, we can infer that increasing the number of sensors or injection rate does not affect X-GSN and JBoss memory usage. In contrast, when the number of sensors is less than 30, the Virtuoso memory usage increases rapidly. However, for the rest of the trial, the Virtuoso memory usage remains steady with some random changes.

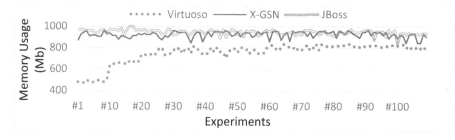

Fig. 6. Memory usage for different experiment settings described in Sect. 3.2 (Color figure online)

In the same format, the CPU usage among all the conducted experiments is shown in Fig. 7. In the case of X-GSN and JBoss, similar to their memory usage, we can observe that increasing the number of sensors or injection rate does not affect their CPU usage percentage. In contrast, when the number of sensors is less than 30, the Virtuoso CPU usage increases rapidly. However, for the rest of the trial, the Virtuoso CPU usage remains steady with some random changes. Full protocols of the described experiments can be found in [17].

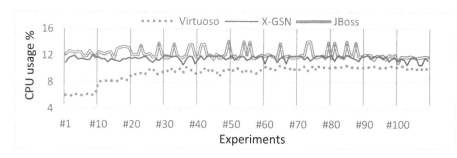

Fig. 7. CPU usage for different experiment settings described in Sect. 3.2

3.5   Analysis of Benchmarking Results

The results presented in Sect. 3.4 has helped establish the baseline performance of OpenIoT using the proposed benchmarking metrics. From the experimental results, it can be noted that the current performance of the OpenIoT platform is suitable for small-scale settings (in-house lab setups). However, when deployed under large scale scenarios, in particular in highly data intensive use cases such as Smart Cities, the performance of the OpenIoT platform will be greatly challenged.

Based on the authors' experience working with OpenIoT, the lack of performance in OpenIoT is mainly due to the imperfect implementation. For instance, the provided implementation of LSM does not support multi-threading or an appropriate buffering mechanism. Therefore, LSM becomes a bottleneck when multiple sensors inject data at a very high rate. This could potentially be overcome introducing some changes to the platform such as adding a queuing system based on ZeroMQ or other such technologies.

Based on the IoT platform study conducted in Sect. 2, we can state that it could be useful to introduce features such as elastic clustering for providing storage scalability, find better options for time-series data, add efficient in-memory caching mechanism and message queue for sustainable interconnection of components during peak loads. However, since this paper was to establish the benchmarking metrics and use this to establish the performance of OpenIoT, further investigation of underlying reason pertaining to lack of performance degradation is out of scope of this paper. Introducing such changes to the OpenIoT platform and evaluating the performance of other platforms to establish the baseline performance of such systems is a part of our future work.

In this section we have performed a set of tests which highlighted some issues of a particular platform. In the following section we discuss the problem of establishing a broader test suite that can be used for benchmarking various IoT platforms under different types of load.

4   IoT Platforms Benchmarking

Comparing the performance of different IoT systems is a complex task. The complexity grows exponentially with the number of features and possible use-cases. However, performance benchmarks are needed for both consumers and developers. Product consumers can rely on the benchmarking results for making a better choice and developers can analyse weaknesses of their product, improve and demonstrate the results. In the world of transactional databases, this effort was started and supported since 1988 by a non-profit organization called Transaction Performance Council (TPC) [18]. Actual benchmarks are TPC-C, TPC-H, TPC-E, TPC-DS, TPC-DI and TPCx-HS which cover such areas as OLTP, ad-hoc DSS, complex OLTP, complex DSS, data integration and Big Data. In the NoSQL movement, which has significant differences in approaches with classical transactional world, the most popular benchmarking approach is the Yahoo Cloud Serving Benchmark (YCSB) [19] which is supported by some open-source tools [20]. Semantic web community also introduces a number of benchmarking strategies for RDF Stores [21]. Regarding OpenIoT, evaluation results for Virtuoso 7 are provided [22] according to Berlin SPARQL Benchmark [23].

Discussion of benchmarking strategies for the IoT platforms has already started [24, 25] and some attempts are already made [2, 26]. For example, HP develops the IoTA-bench [27] with an initial focus on use-cases like smart metering. The problem is in the variety of vendors' understanding of the IoT platforms principles, tasks, main features and system complexity in general. As we have underpinned in Sect. 2, a common way for IoT platforms is to integrate several storage and caching technologies for different types of data. In the dynamic world of IoT, automatic data management strategies are becoming more and more valuable. These data management strategies, supported by the ingestion and retrieval pipelines will affect benchmarking results at the same level as the underlying storage technologies.

5 Conclusion

In this paper, we proposed benchmarking metrics in order to establish the baseline performance of the storage component of IoT platforms by studying and comparing several widely discussed commercial and research IoT platforms. We tested the developed benchmarking metrics using the widely used experimental IoT platform OpenIoT. We conduced several experiments in order to benchmark the performance of the OpenIoT platform. Results of the evaluation show that the bottleneck occurs on the data ingestion and storage components of OpenIoT (namely LSM). While having a great research promise and pioneering contribution to semantic interoperability of IoT silos, OpenIoT could be suitable as an experimental platform for small-scale testbeds. However, experiment outcomes demonstrate that Smart City scale data will pose significant challenges and stress on OpenIoT scalability, reliability and performance. In summary, based on our established benchmarking metrics and experimental evaluation of OpenIoT based on these metrics, we conclude, the currently open source release of OpenIoT as available in GitHub is not ready for substantial large scale commercial IoT deployments without significantly upgrading its performance, reliability, and stability.

As future work, we discussed the need for developing benchmarking standards for IoT platforms. As the area of IoT middleware is comparatively new, the benchmarking methodologies are not well-developed yet. Development of such methodologies is a challenging task due to the complexity of IoT middleware systems and diversity of their components and tasks. However, development of such benchmarks will significantly contribute to the field and bring some amount of understanding to the integrators.

Acknowledgement. Part of this work has been carried out in the scope of the project bIoTope which is co-funded by the European Commission under Horizon-2020 program, contract number H2020-ICT-2015/688203 – bIoTope. The research has been carried out with the financial support of the Ministry of Education and Science of the Russian Federation under grant agreement RFMEFI58716X0031.

References

1. Velosa, A., Natis, Y.V., Pezzini, M., Lheureux, B.J., Goodness, E.: Gartner's Market Guide for IoT Platforms (2015)
2. Vandikas, K., Tsiatsis, V.: Performance evaluation of an IoT platform. In: 2014 Eighth International Conference on Next Generation Mobile Apps, Services and Technologies, pp. 141–146. IEEE (2014)
3. Predix developer network, services and software. https://www.predix.io/catalog/services/
4. Predix Architecture. https://www.predix.com/sites/default/files/ge-predix-architecture-r092615.pdf
5. Tibbo Aggregate IoT Integration platform. http://aggregate.tibbo.com/

6. AggreGate Performance and Scalability Facts. http://aggregate.tibbo.com/technology/architecture/performance.html
7. ThingWorx IoT Technology Platform. https://www.thingworx.com/platforms/
8. Amazon Kinesis. https://aws.amazon.com/kinesis/
9. Amazon AWS IoT. http://docs.aws.amazon.com/iot/latest/developerguide/what-is-aws-iot.html
10. Foster, A.: Enhanced data storage capabilities for IBM Watson IoT Platform. https://developer.ibm.com/iotplatform/2016/07/25/enhanced-data-storage-capabilities-for-ibm-watson-iot-platform/
11. Moltchanov, B., Rocha, O.R.: Generic enablers concept and two implementations for European future internet test-bed. In: 2014 International Conference on Computing, Management and Telecommunications (ComManTel), pp. 304–308. IEEE (2014)
12. Ramparany, F., Marquez, F.G., Soriano, J., Elsaleh, T.: Handling smart environment devices, data and services at the semantic level with the FI-WARE core platform. In: 2014 IEEE International Conference on Big Data (Big Data), pp. 14–20. IEEE (2014)
13. FIWARE semantic application support generic enabler. https://forge.fiware.org/plugins/mediawiki/wiki/fiware/index.php/Semantic_Application_Support_-_Users_and_Programmers_Guide
14. Serrano, M., Quoc, H.N.M., Le Phuoc, D., Hauswirth, M., Soldatos, J., Kefalakis, N., Jayaraman, P.P., Zaslavsky, A.: Defining the stack for service delivery models and interoperability in the internet of things: a practical case with OpenIoT-VDK. IEEE J. Sel. Areas Commun. 33, 676–689 (2015)
15. Aberer, K., Hauswirth, M., Salehi, A.: A middleware for fast and flexible sensor network deployment. In: Proceedings of 32nd International Conference on Very Large Data Bases, pp. 1199–1202 (2006)
16. Le-Phuoc, D., Nguyen-Mau, H.Q., Parreira, J.X., Hauswirth, M.: A middleware framework for scalable management of linked streams. Web Semant. Sci. Serv. Agents World Wide Web 16, 42–51 (2012)
17. Hassani, A.: OpenIoT evaluations. https://github.com/ahas36/openiot/tree/Evaluation/evaluations
18. Transaction Processing Performance Council. http://www.tpc.org/
19. Cooper, B.F., Silberstein, A., Tam, E., Ramakrishnan, R., Sears, R.: Benchmarking cloud serving systems with YCSB. In: Proceedings of the 1st ACM Symposium on Cloud Computing - SoCC 2010, p. 143. ACM Press, New York (2010)
20. Yahoo! Cloud Serving Benchmark (YCSB) github page. https://github.com/brianfrankcooper/YCSB/wiki
21. RDF Store Benchmarking. https://www.w3.org/wiki/RdfStoreBenchmarking
22. Virtuoso BSBM V3.1 Results, April 2013. http://wifo5-03.informatik.uni-mannheim.de/bizer/berlinsparqlbenchmark/results/V7/index.html#exploreVirtuoso
23. Bizer, C., Schultz, A., Pan, Z., Heflin, J.: Berlin SPARQL Benchmark (BSBM) Specification - V3.1. http://wifo5-03.informatik.uni-mannheim.de/bizer/berlinsparqlbenchmark/spec/index.html
24. Malim, G.: Looking for a benchmarking framework for IoT platforms. http://www.iotglobalnetwork.com/iotdir/2016/02/16/looking-for-a-benchmarking-framework-for-iot-platforms-1031/

25. Nambiar, R.: Benchmarking internet of things (CISCO). http://blogs.cisco.com/datacenter/industry-standards-for-benchmarking-iot
26. PROBE-IT benchmarking framework. http://www.probe-it.eu/?page_id=1036
27. Arlitt, M., Marwah, M., Bellala, G., Shah, A., Healey, J., Vandiver, B.: IoTAbench. In: Proceedings of the 6th ACM/SPEC International Conference on Performance Engineering - ICPE 2015, pp. 133–144. ACM Press, New York (2015)

Apps for Environments: Running Interoperable Apps in Smart Environments with the meSchup IoT Platform

Thomas Kubitza[✉]

University of Stuttgart, Stuttgart, Germany
thomas.kubitza@vis.uni-stuttgart.de

Abstract. With *Apps* a popular concept was introduced allowing end-users to easily extend their devices such as smartphones or computers with specific functionality. Two million Apps have ever since found their way into each of the popular App-stores Google Play and Apple Store. We argue that the App-concept is not only well applicable to single devices but also to complete environments equipped with smart networked things. In the moment when Apps can be easily downloaded and executed in home, office and industry environments a wide new applications space will be opened up. In this work we introduce the concept of *Smart Space Apps* that can be downloaded from a cloud-based App-store into a smart environment where they dynamically utilize the capabilities of available smart things to optimally achieve the purpose they were installed for. We introduce a unified schema for the access of sensors and actuators of heterogeneous devices from within Smart Space Apps and describe the middleware and runtime that implements this approach. We explain how Apps are packaged into an exchangeable format and published within a cloud-based App-store. Multiple application use cases are shown and challenges of this novel approach are discussed.

Keywords: Internet of Things · Smart environments · Middleware · Smart Space Apps

With the increasing number of smart devices, networked sensors and programmable actuators many novel opportunities arise through their smart *composition*. Internet of Things (IoT) technologies and networked wearable devices provide new opportunities for creating distributed tangible user interfaces and intelligent behaviour of networked devices sharing the same physical space. For instance, an activity tracker worn by many users today is mainly collecting activity-data throughout the day, however, when its user is sitting on the couch in front of the TV the embedded accelerometer of the activity tracker *could* be exploited to detect arm gestures that control the TV.

As soon as individual smart things are able to offer some or all of their *capabilities* to the smart environment in which they are located a multitude of new useful applications will arise that optimally make use of this distributed sensors, actuators, input devices and screens to assist users during their onsite activities and tasks. At the same time we believe that the concept of *Apps*, which is a popular approach allowing users to easily extend the functionality of single devices, can be also applied to smart environments.

© Springer International Publishing AG 2017
I. Podnar Žarko et al. (Eds.): InterOSS-IoT 2016, LNCS 10218, pp. 158–172, 2017.
DOI: 10.1007/978-3-319-56877-5_10

In this work we report from our ongoing research on realizing this vision. With Apps for Environments we introduce a concept, infrastructure and implementation in which exchangeable Apps that users can download into their smart environments can dynamically utilize available capabilities of networked devices to interact with onsite users and to optimally achieve their purpose.

1 Smart Space Apps Concept

Apps for Environments or *Smart Space Apps* do not run on single devices but instead in environments. For this purpose our approach requires one (stationary) computing node in the local network to run a *middleware* and *runtime* for the execution of smart space Apps. This node can be part of an existing networking infrastructure (e.g. WiFi, Ethernet) or span its own networks using different communication technology adapters (smart hub). A main pillar of our Smart Space Apps concept is a *unified view* on networked computing devices and their *capabilities*: In general, if a device has a processor and a communication interface it can become part of our smart space. Whether certain capabilities of a device should be exposed to the smart environment or not is a decision that inhabitants of a smart space have to decide on inclusion of a new device to the network. Applied to the initial example - the user that controls his TV using his activity trackers arm band, has at some point decided to expose his trackers accelerometer to be used on demand by his domestic environment. In addition he has downloaded and installed a Smart Space App into his smart hubs runtime that promises to utilize devices of the type "activity-tracker" to control devices of the type "TV". After download and execution the functionality is instantly available. The core functionality of the users' activity tracker, measuring and reporting activity, is still continued but in addition sensor-data events measured by the accelerometer are shared with the users' smart environment as soon as the tracker is in its wireless range.

Unified Access Schema
Our unified view on capabilities of smart things breaks these down into *sensors* (devices that are pure data producers), *actuators* (devices that are pure command receivers) and combinations of both. This generic view allows to integrate networked devices of very different kinds in the same way: Capabilities of commercial IoT devices, networked DIY sensors, smart phones, tablets or home appliances can be accessed in the same way. This forms the basis of a *unified access schema* that is used within our runtime to give Apps access to the shared capabilities of an environment. This schema is illustrated in Fig. 1.

This schema uses a (locally) unique user-defined or auto-generated name to reference specific devices, device-wide unique names for modules that represent sensors, actuators, or combinations and names for properties of the referenced module to access sensor data or to trigger actuator commands. As indicated by the dot-notation a runtime could inject this hierarchically structured information into its namespace and execute Smart Space App code written in any programming language that uses this access schema to implement behaviour that interweaves sensor events with system state and actuators. Using the devices indicated in Fig. 1 a minimal Smart Space App could be implemented

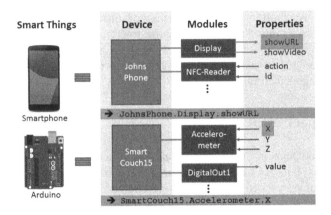

Fig. 1. Generic modularization approach and naming scheme used for referencing smart things and their capabilities

with the following code that shows the weather forecast on the display of "JohnPhone" when the accelerometer of "SmartCouch15" is actuated (e.g. when someone sits down).

```
if (SmartCouch15.Accelerometer.X > 10) {
    JohnsPhone.Display.showURL ("http://bing.com?q=weather");
}
```

This very simplistic example illustrates the conceptual level on which logic for Smart Space Apps is implemented. Note that the networking layer is completely abstracted and that devices and capabilities of entirely different platforms and operation system are accessed in the same way. Although very basic, these three lines of code already implement a small Smart Space App that interweaves two specific devices. Complex Apps can consist of thousands of lines of code, interweaving arbitrary numbers of devices, sensors and actuators. In contrast to the example above Smart Space Apps that should run not just in one specific environment, typically do not consist of code that references devices *specifically* (e.g. "JohnsPhone") but instead reference devices dynamically by their type and capabilities (see section "Runtime" for an example). This allows building and publishing Smart Space Apps that can be downloaded and run in smart environments that consist of very different devices than the ones they were originally build and tested in. For instance a Smart Space App that notifies users when movement sensors were triggered could use flashing the floor-lights in smart space A while in smart space B the same App would show (in addition) a notification message on the displays of currently near tablet or smartphone devices.

In the previous paragraphs we explained the concept behind Smart Space Apps. In the next section we will briefly sketch the implementation of this concept in the form of the meSchup IoT platform.

2   meSchup IoT Platform

The *meSchup IoT platform* [1, 2] was designed and implemented within the four year FP7 EU project *meSch*. meSchup consists of an integrated *middleware, runtime* and web-based *development environment (IDE)* software for onsite hub computers as well as client software, firmware and adapters for a wide range of client devices. The client support at the time of writing incudes Android devices (smartphones, tablets, projectors, TV, Amazon Fire) smart things platforms such as Arduino, Espressif, .NET Gadgeteer and nRF51822 microcontrollers, Raspberry Pi and Intel Edison computers, Windows and Linux machines, smart-plugs, ambi-lights and multiple other smart appliances. For devices that by default are not able to expose their capabilities to local smart environment Apps, client software or firmware is provided. For devices that are not extendable with Apps or firmware such as typical smart X appliances adapters are offered that run in the meSchup middleware and allow direct communication or control of these devices. The meSchup server software is fully implemented in Node.js and can be thus run platform independent on various operation systems. Although meSchup can run on any off-the-shelf machine we provide a set of different meSchHub devices that come optimally preconfigured for various purposes. Figure 2a for instance depicts the meSchup proto-typing hub (meSchHub-P) that is based on a Raspberry Pi II computer and comes with its own embedded WiFi Hotspot, battery and meSchup platform preinstalled on SD-card.

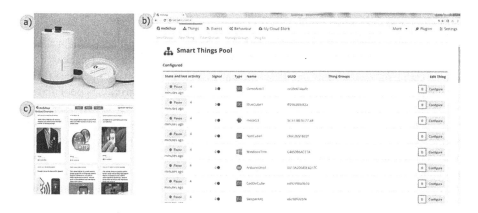

Fig. 2. (a) meSchHub-P device, (b) meSchup web interface: Overview of configured and recently discovered smart things, (c) Public Smart Space Apps in the cloud-based App store

Middleware
The middleware layer of the meSchup platform abstracts from arbitrary (wireless) communication technologies and protocols and provides unified bidirectional access to remote devices and their capabilities for higher layers of meSchup such as the App-runtime, GUI or IDE. Adapters for various communication technologies are supported such as WiFi, ZigBee, Z-Wave, BLE and LoRa. For each of these communication technologies automatic device-discovery is implemented. Discovered devices are listed in

the web-based smart things pool overview where users can configure which of their capabilities they want to provide to their smart environment (Fig. 2). Known devices are automatically discovered and configured every time they return into the (wireless) network. This in particular allows roaming devices such as smartphones to be alternately used in different smart environments. Event-based communication is the default communication model between meSchup clients and the middleware server. This means that client devices do the "hard job" locally such as fast sampling of sensors or complex computation. Only when adequate sensor changes are detected clients will send sensor events to the meSchup middleware. This will trigger Apps to be executed in the meSchup runtime. However, other behaviours such as time-series data can be also configured. Instant remote reconfiguration allows to expose new sensors and actuators anytime without the need to reinstall a client or re-flash firmware of a smart device. This is particularly important to minimize the maintenance for potentially hundreds of devices that are expected in near future smart environments [3]. The event-based communication model saves wireless bandwidth, keeps communication responsive and scales for large numbers of devices.

The meSchup middleware is particularly designed to be extendable and strongly inclusive. New communication technologies and protocols can be easily added as middleware modules. By default popular IoT protocols such as MQTT and HTTP/REST are supported by the platform. Virtual devices optimally map the sensor- and actuator-topics/resources of these protocols onto the unified device-module-property access schema.

Web-Based User Interface

meSchup comes with its own web-based GUI for device- and App-management as well as an integrated development environment for the creation, testing and packaging of Smart Space Apps. After starting the meSchup software or powering up a meSchHub the interface is instantly accessible via "http://meschup". The smart things pool view of the UI gives an overview of currently available smart things (Fig. 2b), their connection status, configuration and new devices that have recently been discovered. New sensor/actuator modules can be added or removed in a drag&drop manner and are made instantly available on the remote device. The events view allows monitoring of sensor and actuator events for debugging purposes. The behaviour view provides an overview of currently installed Smart Space Apps and allows to enable/disable, remove or edit Apps (and their underlying interaction scripts). Foreign Apps can be downloaded from a cloud-based Smart Space App store and new Apps can be developed within the integrated IDE and be subsequently published to the store (Fig. 3).

Runtime

The meSchup runtime uses *JavaScript* as the programming language for the implementation of Smart Space Apps. An App can consist of one or many *interaction scripts*. The previously introduced unified access schema is exposed as the object api.device within the runtime and provides all Apps unified access to all devices and capabilities of the smart things pool. An exemplary interaction script is shown below. This generic

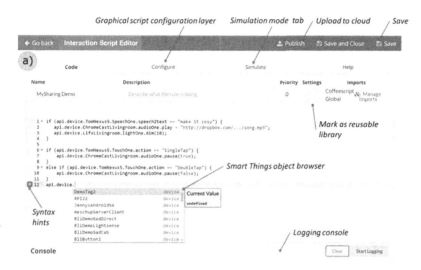

Fig. 3. Integrated web based development environment (IDE) for Smart Space Apps

interaction script displays the incoming message on any device that offers a *display-capability* to the local environment.

```
 5   ...
 6   // Iterate over all smart things
 7 ▾ for (var device in api.device) {
 8       // Show msg on screen if device offers a display
 9 ▾     if (hasModule(device,"display")) {
10           api.device[device]["display"].showHtml ("<b>Msg for John:</b>"+msg);
11       }
12   }
```

Interaction scripts are executed event-based when new sensor events are received by the middleware or timer-based to react on schedules. The developers of App-interaction scripts decide on which events to listen. However, this can be restricted on installation of Apps by not granting access to certain device or module types. The runtime allows executing many Smart Space Apps in parallel. This makes it easy to extend environments continuously with specific functionality required by their inhabitants.

A Smart Space App can optionally provide a graphical configuration user interface that allows End-users to fine-tune its behaviour to their needs and local environment without requiring programming knowledge. These interfaces are web-based and will be typically accessed via users' smartphones or tablets. Developers of Smart Space Apps decide how extensive configuration options are and whether they want to provide a configuration interface at all.

App Store

Smart Space Apps package interaction scripts and optional resources that are for instance required for configuration UIs (HTML, CSS, images, etc.) together with meta-data into exchangeable compressed Smart Space Apps files (.S2A file extension). Meta information includes among others the App-title, description, tags, author and version. Further, information on necessary device types and modules is included. The packaging process is integrated into the web-based IDE. Packaged S2A-Apps can be uploaded into our cloud based App store where they can be easily found by interested users, if the author decides to make them public. The App store requires a previous registration of a user account. This account can be then assigned to one or many meSchHubs that are managed by the same user or organisation. Packaged Apps are then automatically uploaded to this cloud-based App store account but are not made public. Users can either decide to make their Apps publicly available or to keep them private. Users or organisations managing multiple meSchHubs can use the App store to remotely deploy new or updated Apps to one or many of their hubs at once. We currently extend this basic functionality of our App store with additional community functions such as author- and App-ratings, comments, bug-reports and feature requests.

3   Example Applications

meSchup is used as core IoT platform for realizing novel IoT applications in multiple different projects and domains. Three application examples from these projects are briefly described to indicate the broad range of current and future potential application areas for Smart Space Apps.

Smart Interactive Exhibitions
meSchup is actively used in multiple European museums and cultural heritage institutions as the core onsite infrastructure for the realisation of smart interactive exhibitions. Multiple of these exhibitions were packaged up and are available in an App store[1] that was particularly designed for cultural heritage professionals (CHPs). This App store conceptually works in a similar way as the previously described generic App stores but in addition it provides support for the typical workflows of exhibitions designers. In particular it provides easy access to content sources for CHPs such as Europeana[2] and allows to interweave this content conveniently with the smart behaviour of available smart exhibition Apps (these are called *recipes* within mesch.io).

The example application depicted in Fig. 4 for instance is deployed across multiple points of interest (POIs) within a museum exhibition and allows a visitor to perceive personalized multi-media content at each POI based on an object that the visitor has picked at the entrance. This object represents a perspective on the exhibition (e.g. in the context of a First-World-War exhibition narrations from the perspective of a soldier versus the one of a civilian) and the chosen language of the visitor. At each POI a pulsing round area indicates that visitor can place objects on it. This explicit interaction triggers audio, video or other presentations that are tailored to the visitors' language and perspective.

Technically this interactive setup is realized with NFC readers, projectors, screens and earpieces embedded into each POI and NFC tags embedded into the wearable objects for visitors. Apps running in the onsite meSchup platform interweave events from NFC

---

[1]   http://mesch.io.

[2]   http://www.europeana.eu/portal/en.

Fig. 4. This point of interest in an interactive exhibition space provides personalized multi-media content using a personal NFC-based token that visitors use across an exhibition

readers with the appropriate content and instantly trigger the display of media at the corresponding points of interest. Associations of content to POIs and NFC objects are not hardcoded but instead use simple semantic tags to derive which content should be shown in which situations. Besides implementing this visible behaviour for visitors the responsible App can also collect anonymised usage data and provide spatial and temporal usage-visualisations for curators.

This example illustrates how meSchup and the Smart Space App concept facilitate the realisation of advanced distributed, customizable interactive installations for novel museum experiences without requiring any low-level programming experience from museum-curators and exhibition designers. Furthermore, installations realized in one physical setup can be easily transferred into another physical setup by just installing the same App on another smart hub. Full interactive exhibition floors can completely switch their purpose and content by simply installing another Smart Space App.

Ubiquitous Notifications

meSchup is also utilized in research projects that explore ubiquitous notifications in the context of smart home and office environments as well as ambient assisted living. In the ubiquitous notifications project [4] meSchup is used to display notifications that arrive at users smartphones (e.g. WhatsApp, Email, Calendar, etc.) instantly in the users environment by dynamically using the capabilities that the current devices in proximity offer. Notifications can thus be overlaid on a TV screen in one situation while they are read aloud by an Amazon Echo device or indicated with ambient lights in another situation. Multiple mobile and stationary deployed sensor sources such as BLE beacons, movement sensors and phone status and orientation are utilized to optimally derive the best devices and output modalities for displaying notifications. Different Smart Space Apps provide an easy way to deploy and test different visualisation concepts and strategies in different physical setups and with different users (Fig. 5).

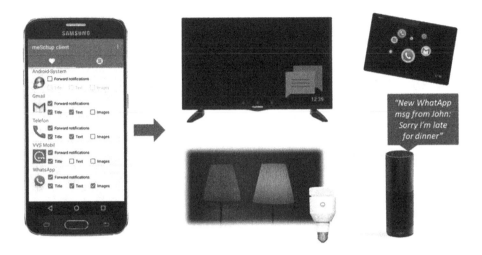

Fig. 5. Ubiquitous notifications: Notifications received on users' smartphones can be instantly forwarded to smart devices in a users proximity, such as TVs and screens, ambient lights, photo-frames and tablets as well as speakers (e.g. Amazon Echo).

In a similar approach the DAAN project[3] utilizes meSchup to realize scenarios for ambient assistive living environments in which contextual information from sensors in the environment are continuously collected and analysed in order to provide ambient proactive suggestions for assistance and behaviour change [5].

Extension of Conventional Things with Smart Interfaces

meSchups instant access to different capabilities of smart things in the same environment allows flexible combinations of sensors, input devices and actuators that physically belong to separate devices. This allows using one or multiple input devices spontane-ously to control actuators of other devices. Associations between input and output can be dynamically established using for instance physical tokens such as NFC tags, BLE tags or visual markers. For instance lying down a smartphone on an NFC tag placed in a smart living room could present a dynamically generated GUI for controlling living-room devices. Using a camera and display of a smartphone in combination with markers placed on various devices could overlay these with new or additional interfaces. Figure 6 for instance shows an example in which a small remotely controllable lamp is extended with an Augmented Reality (AR) switch interface that allows turning the lamp on or off.

The concept of such smart interfaces is applicable to many different domains and scenarios: In industrial settings conventional physical instruments (e.g. temperature and pressure gauges) can be augmented with additional data or functionality. For instance overlays of historical graphs (e.g. the temperature curve of the last 24 h) can be shown in addition to the current gauge-value. In medical environments touchless interaction via AR interfaces can help to keep devices sterile. The shown AR interfaces example is

---

[3] http://daan.dfki.de/.

Fig. 6. Augmented Reality Interface extending a smart mobile lamp with a virtual on/off switch. Pressing the virtual switch on the screen instantly turns the lamp on or off.

realised using a Smart Space App that dynamically overlays identified visual markers with graphical interfaces (images). These are chosen depending on the capabilities of the device that is associated with the marker. Overlays are packed with Smart Space Apps and are thus easy to update and to distribute.

4   Related Work

meSchups' underlying concept lies at the heart of traditional Ubicomp research and was inspired by many of its former research projects.

On a device level a set of platforms have laid the way for simplifying the creation of **sensor actuator equipped smart things**. Smart-its [6] have early allowed experimentation with sensors and actuators in academic research. As one of the first commercially available physical prototyping toolkits Phidgets [7, 8] enabled access to electronic components through a pluggable hardware design and easy to use libraries. D.tools [9] allowed designers to iteratively create physical UIs using a state-chart based programming model. These were followed by the .NET Gadgeteer platform [10] with it solderless pluggable module design and powerful Visual Studio IDE as well as the Arduino [11] platform with its simple to setup code editor and huge community support. The "App" concept introduced with Symbian feature-phones and continued by iOS and Android finally also opened up phones to be used as sensor actuator rich platforms that can run user defined code. These systems and platforms facilitated the creation of *standalone* smart devices by simplifying the development of embedded software and the access to sensor and actuator hardware.

meSchup's concept on device level differs from these previous approaches by offering generic ready-to-go firmware/client-software/Apps for all these device platforms instead of supporting developers to build custom firmware for each and every thing. The purpose of this generic firmware is to make the device and its capabilities (sensors and actuators) accessible through its network interface and to be discoverable and fully controllable by the local smart hub and the Smart Space Apps that run on it.

This approach shifts the individual high level application logic into the hubs of an environment and provides unified access to the *capabilities* of heterogeneous devices independent of their communication technology, used communication protocol or device platform. The generic firmware already handles all platform specific low level calls, local sensor sampling as well as the secure transport of messages from or to a device. Sensor changes of connected devices simply appear as events within the App runtime of the smart hub and actuator commands are simple function calls within an App. This results in a drastically reduced time and effort for developing distributed logic for heterogeneous IoT devices.

The realization of **distributed multi-device interaction** was addressed by a multitude of systems in particular in the domain of context aware computing and smart spaces. iStuff [12] allowed the prototypic exploration of various novel ubicomp scenarios providing easy means for connecting input and output of distributed devices. Dey et al.'s. framework for prototyping context-aware applications [13] laid the foundations for further projects targeting at enabling end-users. iCap [14] provided a pen based interface that allowed end-users to realise individual context aware applications by graphically composing simple event-condition-action rules. Among others centralized architectures were also proposed in iStuff [12] and SEAP [15]. However, these used rule-based languages that were limited in their expressiveness and complexity. More recently some projects have also picked up JavaScript for executing inter-device behaviour. Fabryq [16] supports mobile scenarios using smartphones as gateways to some BLE devices and hosts the JavaScript based application logic in the cloud. Weave [17] focuses on simplifying the synchronisation of GUIs across multiple device displays.

While most of these projects represent proof-of-concept implementations or toolkits for prototyping that cannot be easily transferred out of the lab, meSchup stands for a highly modular generic IoT platform that provides a wide support for available IoT device platforms and high flexibility through its App based runtime. Its capabilities of providing support from local App-development, to App-packaging, App-Store upload and deployment and execution of the same App on other smart hubs is unique among current IoT systems and research projects.

Multiple cloud based platforms exist that offer end-users simple to use interfaces for the realisation of trivial IoT scenarios. The IFTTT service [18] for instance offers a form based web UI that allows users to compose simple trigger-action rules binding one event to one action. Slightly more advanced interconnections between data sources and data sinks can be created with the flow-based visual programming interface of the Node-Red toolkit [19]. However, for more complex applications the graphical flow based UI expands quickly in space and becomes hard to handle.

In contrast to purely cloud based approaches meSchup by design executes its application logic locally in its runtime and is thus robust against internet connection problems and high latencies. Further, the owner of a smart hub has full control over the data collected through Apps and is not forced to send any data to external parties. meSchup Apps written in JavaScript can realize anything from simplest IFTTT rules using a single sensor and an actuator to complex applications that utilize the full power of the JavaScript language while interacting with hundreds of sensors and actuators. Optionally meSchup Apps can push visual GUIs to devices with displays such as smartphones and tablets. These dynamically generated or static GUIs can be used as configuration layer that makes Apps adjustable to end-users without programming skills. Using pure web-technology for Smart Space Apps is a design decision that allows the huge community of web-developers to instantly start the development of IoT applications.

5 Discussion

The concept of *Apps for Environments* for the first time brings easy extendibility with new functionality to smart multi-device environments in the same way as Apps extend a smartphones functionality. This is achieved by providing a unified interface and an access schema that can map and address arbitrary sensors and actuators of heterogeneous remote devices in a unified way. In a similar way operation systems such as Android or iOS provide unified APIs to control the capabilities of various devices and hardware types using the same App. However, smart environments differ in their exponentially higher potential heterogeneity. All devices in a smart environment as well as their capabilities can be used by multiple Smart Space Apps in parallel. This can in particular result in conflicts where the same actuator resource is manipulated by multiple control sources which can lead to unexpected behaviour. Conflict detection and resolution mechanisms are required to handle such situations. MeSchups' UI offers a specific monitoring view in which such cases are detected and indicated by the runtime and can be resolved by end-users. Our approach allows a user to resolve such conflicts by either deactivating one of the conflicting Apps completely or by specifically withdrawing one of both Apps the right to control the conflicting actuator.

Another challenge in smart environments is the omnipresent disruption of connectivity through wireless connection loss or battery drain. Mobile devices can physically get out of wireless range, be interfered by other signals or suffer high delay through low bandwidth or high traffic. A smart environment needs to be able to detect such situations, handle situations in which devices become unavailable and recover devices and their actuator states as soon as they return. meSchup handles situations in which devices become unavailable pragmatically by dropping all actuator commands that are triggered by Apps as long as the target device is not available. App developers can also implement special event-handlers that are executed when devices become available again, allowing for individual initial configurations for certain device types.

We believe that the most challenging aspects in relation to smart environment and Smart Space Apps are privacy and security. Smart environments offer a multitude of new data sources to collect, analyse and derive personal information about their inhabitants. In contrast to cloud based approaches for controlling smart environments (e.g. IFTTT[4]) the meSchup platform offers data privacy by design because it does not require any communication with external servers. Whether Smart Space Apps can communicate with external sources needs to be fully transparent and controllable by smart space inhabitants to assure the privacy of their data. In meSchup users who install new Smart Space Apps have the opportunity to restrict the access to resource-types of the local

---

[4] https://ifttt.com/.

smart space as well as to the internet (by default blocked). Currently meSchup offers individual white and blacklists for device types. However, our experience from various deployments indicates that capabilities of smart devices should in the future be annotated with a metric for privacy and security criticality. For instance doorlocks, cameras and microphones would fall into a higher criticality group than for instance switches or lamps. Such groups would then make it easier for users to decide when granting rights on installation.

Mechanisms are further required to assure the functionality, quality and harmlessness of publicly downloadable Smart Space Apps. Installing untrusted Smart Space Apps potentially contains a multiple of privacy and security risks compared to traditional single-device Apps: Cyber-physical systems such as doors, climate control or heating can be potentially locked, stopped or misused. MeSchups' App packaging allows App creators to sign their App with their certificate, ensuring that unmodified Apps can be installed from trusted parties. Further our App store soon allows the rating and commenting by users, thus indicating the satisfaction with a publicly available App.

To prevent theft of sensible sensor data or spoofing of critical control commands within the networks of smart environments end-to-end encryption is necessary on-top of the existing encryptions mechanisms of the individual communication technologies. MeSchups' middleware addresses communication security in three layers. The first layer involves transport layer security such as WPA for WiFis and password based encryption for BLE and ZigBee. On a second layer architectural security means are applied. For instance smart things only accept messages from the same hub after they have been discovered. On a third layer all communication is in addition encrypted with a per-device-key that has been exchanged on initial inclusion of a smart thing to a smart environment. For instance for Android devices a QR-code is scanned to exchange an initial encryption key by-passing unsecure RF communication (out-of-band). Similarly microcontrollers with installed meSchup firmware and USB/Serial/NFC interface can be attached to the hub for a few seconds to exchange a key securely. Encryption on application layer prevents theft of data and guarantees the identity of previously included devices.

meSchups security roadmap further plans the usage of secure elements base on elliptic curve cryptography (ECC) both on future hub-hardware as well as IoT devices equipped with meSchup firmware. This will simplify the secure initial key exchange among these devices.

6   Conclusion

In this paper we describe the concept of Apps for Environments and its implementation as Smart Space Apps based on the meSchup IoT platform. We introduce a unified schema for accessing capabilities of smart things as the foundation for Smart Space Apps and describe its implementation on top of a middleware and runtime. We explain the implementation of Apps, their packaging into an exchangeable format and the distribution via a cloud based Apps store. We introduce multiple example applications spanning across different domains and present challenges of this novel approach.

We believe that the concept of Apps for Environments has an enormous potential to bring innovative applications into current and future smart environments. Our Smart Space App programming approach provides a clean abstraction from low level layers and is fully based on web-technologies such as JavaScript, HTML and CSS. In combination with the integrated development environment and cloud store support we believe that Smart Space Apps are an attractive platform for a broad range of developers. A wide uptake of this concept would lead to a multitude of new useful IoT applications and creative multi-device solutions.

Acknowledgements. This work is funded by the European Project meSch (Grant Agreement No. 600851).

References

1. Kubitza, T., Schmidt, A.: Towards a toolkit for the rapid creation of smart environments. In: Díaz, P., Pipek, V., Ardito, C., Jensen, C., Aedo, I., Boden, A. (eds.) IS-EUD 2015. LNCS, vol. 9083, pp. 230–235. Springer, Cham (2015). doi:10.1007/978-3-319-18425-8_21
2. Kubitza, T., Schmidt, A.: Rapid interweaving of smart things with the meSchup IoT platform. In: Proceedings of the 2016 ACM International Joint Conference on Pervasive and Ubiquitous Computing Adjunct - UbiComp 2016, pp. 313–316. ACM Press, New York (2016)
3. van der Meulen, R., Rivera, J.: Gartner says a typical family home could contain more than 500 smart devices by 2022. http://www.gartner.com/newsroom/id/2839717
4. Kubitza, T., Voit, A., Weber, D., Schmidt, A.: An IoT infrastructure for ubiquitous notifications in intelligent living environments. In: Proceedings of the 2016 ACM International Joint Conference on Pervasive and Ubiquitous Computing Adjunct - UbiComp 2016, pp. 1536–1541. ACM Press, New York (2016)
5. Wiehr, F., Voit, A., Weber, D., Gehring, S., Witte, C., Kärcher, D., Henze, N., Krüger, A.: Challenges in designing and implementing adaptive ambient notification environments. In: Proceedings of the 2016 ACM International Joint Conference on Pervasive and Ubiquitous Computing Adjunct - UbiComp 2016, pp. 1578–1583. ACM Press, New York (2016)
6. Beigl, M., Gellersen, H.: Smart-its: an embedded platform for smart objects. In: Smart Objects Conference (2003)
7. Greenberg, S., Fitchett, C.: Phidgets: incorporating physical devices into the interface. In: Proceedings of UIST 2001, pp. 209–218. ACM Press (2001)
8. Greenberg, S., Fitchett, C.: Phidgets: easy development of physical interfaces through physical widgets. In: Proceedings of the 14th annual ACM symposium on User interface software and technology - UIST 2001, p. 209. ACM Press, New York (2001)
9. Hartmann, B., Klemmer, S., Bernstein, M.: d.tools: integrated prototyping for physical interaction design. In: IEEE Pervasive Computing (2005)
10. Villar, N., Scott, J., Hodges, S.: Prototyping with microsoft .net gadgeteer. In: Proceedings of the Fifth International Conference on Tangible, Embedded, and Embodied Interaction - TEI 2011, p. 377. ACM Press, New York (2011)
11. Arduino: Physical prototyping platform. https://www.arduino.cc
12. Ballagas, R., Ringel, M., Stone, M., Borchers, J.: iStuff: a physical user interface toolkit for ubiquitous computing environments. In: Proceedings of the conference on Human factors in computing systems - CHI 2003, p. 537. ACM Press, New York (2003)

13. Dey, A., Abowd, G., Salber, D.: A conceptual framework and a toolkit for supporting the rapid prototyping of context-aware applications. Hum. Comput. Interact. **16**, 97–166 (2001)

14. Dey, A.K., Sohn, T., Streng, S., Kodama, J.: iCAP: interactive prototyping of context-aware applications. In: Fishkin, K.P., Schiele, B., Nixon, P., Quigley, A. (eds.) Pervasive 2006. LNCS, vol. 3968, pp. 254–271. Springer, Heidelberg (2006). doi:10.1007/11748625_16

15. Holloway, S., Stovall, D., Lara-Garduno, J., Julien, C.: Opening pervasive computing to the masses using the SEAP middleware. In: 2009 IEEE International Conference on Pervasive Computing and Communications, pp. 1–5. IEEE (2009)

16. McGrath, W., Etemadi, M., Roy, S., Hartmann, B.: Fabryq. In: Proceedings of the 7th ACM SIGCHI Symposium on Engineering Interactive Computing Systems - EICS 2015, pp. 164–173. ACM Press, New York (2015)

17. Chi, P.P., Li, Y.: Weave: scripting cross-device wearable interaction. In: Proceedings of the 33rd Annual ACM Conference on Human Factors in Computing Systems - CHI 2015, pp. 3923–3932. ACM Press, New York (2015)

18. IFTTT: 'If this then that' cloud service. https://ifttt.com

19. Node-Red: A visual tool for wiring the Internet of Things. http://nodered.org

Semantic Interoperability at Big-Data Scale with the open62541 OPC UA Implementation

Julius Pfrommer[(✉)]

Fraunhofer IOSB, Fraunhoferstraße 1, 76131 Karlsruhe, Germany
`julius.pfrommer@iosb.fraunhofer.de`

Abstract. The OPC Unified Architecture (OPC UA) is a protocol for Ethernet-based communication in industrial settings. At its core, OPC UA defines a set of services for interaction with a server-side information model that combines object-orientation with semantic technologies. Additional companion specifications use the OPC UA meta-model to define domain-specific modeling concepts for semantic interoperability. The *open62541* project is an open source implementation of the OPC UA standard. In this work, we give a short introduction to the core concepts of OPC UA and how the measures taken to scale OPC UA to Big-Data scale reflect in the architecture of *open62541*.

Keywords: OPC UA · Open source · Semantic interoperability · Big-data

1 Introduction

In the past, a multitude of communication technologies and protocols have been used for data exchange in industrial settings. The reasons for this are the diverse use cases and their requirements, such as realtime communication in safety-critical applications, but also a lack of interoperability between vendors. Today, it has become common to deploy Ethernet-based networking in addition to traditional fieldbus-based communication systems [6]. With this increase in flexibility for communication, new applications have emerged. For example Plug&Produce [1,16], where system components are equipped with a self-description for fast deployment and configuration, and Condition Monitoring [5,14], where analytics based on runtime data is used to improve operations and especially maintenance.

The OPC UA protocol promises not only to unify industrial communication but also to enable semantic interoperability. Motivated by the increased use of Big-Data concepts in the industry, this work discusses how OPC UA can be scaled up to support large data volume, velocity and variety. The remainder of this paper is structured as follows. The OPC UA protocol and its capabilities for information modeling and semantic interoperability are introduced in Sect. 2. Section 3 discusses the use of OPC UA in Big-Data scenarios. In light of the features described in the previous two sections, we motivate the design of the *open62541* OPC UA implementation and the measures taken to ensure scalability in Sect. 4. The paper concludes with a summary in Sect. 5.

© Springer International Publishing AG 2017
I. Podnar Žarko et al. (Eds.): InterOSS-IoT 2016, LNCS 10218, pp. 173–185, 2017.
DOI: 10.1007/978-3-319-56877-5_11

2   Semantic Interoperability with OPC UA

According to Heiler [8], interoperability in distributed systems is the ability to exchange data and use remote services based on agreements for message passing protocols, procedure names, error codes, and so on. Semantic interoperability additionally requires a common understanding of the meaning of the requested services and data. In this section, we give an introduction to the capabilities of OPC UA for information modeling and the relation to established semantic technologies, such as ontologies [10].

Developed as the successor to the widely used OPC Classic protocol, OPC UA has become a major contender for Ethernet-based communication in industrial applications and has been standardized in IEC 62541 [11]. At its core, OPC UA defines

- a type system to define protocol messages with a binary and XML-based encoding scheme,
- a meta-model for information modeling that combines object orientation with semantic triple-relations, and
- a set of 37 standard services to interact with a server-side information model. The signature of each service is defined as a request and response message in the protocol type system.

The OPC Foundation drives the continuous improvement of the standard, the development of companion specifications, and the adoption of OPC UA in the industry by hosting events and providing the infrastructure and tools for compliance certification.

An OPC UA server exposes its information model to remote clients. In cyberphysical systems, the server is usually deployed close to the source of information, i.e. the physical process. OPC UA information models can be fully introspected at runtime using only the standard services. So a client that connects to a server for the first time needs no prior notion of the server's content. The idea then is to define reusable building blocks for OPC UA information models. This can drastically reduce the effort required for integration when the client software, built with a notion of these reusable building blocks, adjusts at runtime to the remote information model it encounters.

In order to support cross-vendor interoperability, the so-called OPC UA companion specifications map established standards from an application domain to OPC UA information models. That is, companion specification define such reusable building blocks for OPC UA information models (reference types, data types, variable types, methods and object types, see their discussion later in this section) together with their intended use. The current companion specifications are based on established standards in their application domain and are driven by joint working groups between the OPC Foundation and domain-specific standardization bodies or industrial consortia. Existing companion specifications in the manufacturing automation domain are, for example, Field Device Integration, ISA-95, PLCopen, MTConnect and AutomationML.

Table 1. Node classes in OPC UA

Node Class	Usage
ReferenceTypeNode	Predicate type to be used in references between nodes
DataTypeNode	A data type for scalar values
VariableNode	A named value (scalar, array, multi-dimensional array)
VariableTypeNode	Requirements for variables (data type, array size and so on)
MethodNode	A callable remote procedure with its signature
ObjectNode	An object made up of variables, methods and further objects
ObjectTypeNode	Requirements for objects (mandatory and optional members)
ViewNode	Gives access to a subset of nodes

At its core, every OPC UA information model consists of nodes and typed references, each linking two nodes in a directed graph. Every node is of one of the eight node types shown in Table 1. The references can be thought of as *NodeId* tuples of the form (source, predicate, target), where the predicate denotes a *ReferenceTypeNode*. In the remainder of this section, we introduce the different node classes and discuss how they are used as part of an object-oriented information model.

Reference Types. Every reference uses the *NodeId* of a *ReferenceTypeNode* for its predicate. The predefined *ReferenceTypeNodes* shown in Fig. 1 form an extensible hierarchy. Every reference type has three binary properties: hierarchical/nonhierarchical, abstract/concrete, symmetric/directed. Hierarchical reference types are subtypes of *HierarchicalReferences*. They may not form directed cycles. References with a symmetric reference type are undirected (consider e.g. an *isEqualTo* reference). Abstract reference types cannot be used directly and are used to structure the reference types hierarchy.

The following example shows how reference types alone can be used for semantic interoperability. Assume we want to model the layout of a technical system. For this, we introduce two custom reference types. First, the hierarchical *contains* reference type indicates that a component (subsystem) is part of a larger system. This gives rise to a tree of containment relations. For example, a motor is contained in the car and a crankshaft is contained in the motor. Second, the symmetric *connectedTo* reference type indicates that components are connected outside of the containment hierarchy. For example, the car wheels are connected to the axle. A client can then learn the layout of system represented in an OPC UA information model based on a common understanding of just two additional reference types. Further subtypes of *connectedTo* could be used to differentiate between physical, electrical and information related connections.

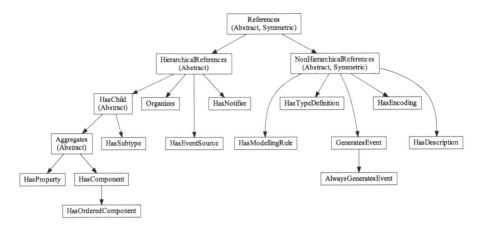

Fig. 1. Predefined reference types in OPC UA. Arrows denote a *hasSubType* reference.

Variables. VariableNodes are used as named value containers similar to their use in programming languages. Variables are typed and store both their current value and the constraints that need to hold the value. The constraints are the data type, the value rank (scalar, array, multi-dimensional array, etc.) and, for multi-dimensional arrays, the array length in each dimensions. VariableNodes may also have child variables. These may give additional information, for example the unit of the value. The semantics of a variable may be inferred from the VariableTypeNode it references.

Variable Types. Naming a variable "speed" may not be enough for cross-vendor interoperability, as speed can be expressed in many ways. For this, variable types are a more suitable choice. Every variable references exactly one *VariableTypeNode*. The variable type both convey meaning and imposes constraints for possible variable values (or rather for the data type, value rank and array dimensions of the variable). Again, variable types form a type hierarchy and can be abstract. Assume now the "speed" variable is of a variable type *FloatingPointRPM*. Both the encoding and the meaning of the variable are then defined in the variable type in enough detail write program code against the abstraction that then applies to all instances.

Data Types. DataTypeNodes define scalar values. Built-in data types comprise for example integers of various size, strings, and so on. Users can add custom data types by combining the built-in data types to form structures, unions and arrays. An example are 3D-coordinates represented by a tuple (x:float, y:float, z:float). The data types also form a hierarchy. Similar to *ReferenceTypesNodes*, abstract data types are used to structure the hierarchy. For example, an *increaseCount* method might refer to the (abstract) *UInteger* data type for unsigned integers in its signature. The actual values sent over the wire are then of a concrete data type derived from *UInteger*, such as *Byte* or *UInt16*.

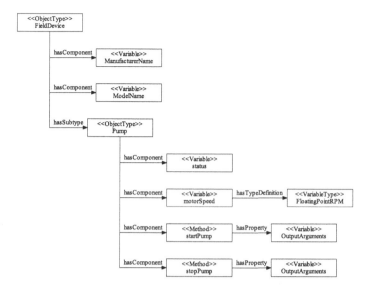

Fig. 2. Excerpt of an object type hierarchy.

Objects. Objects are used to structure information models and contain variables, methods and further objects (i.e., refers to them with a hierarchical reference). For example, in the representation of a car and its technical layout that was just introduced, every component would be represented by an object, including the car itself. For interoperability, it is important to impose structure. Programmable logic controllers (PLCs) in the automation domain have traditionally exposed simply flat lists of variables in their interfaces. Much manual effort was needed to define which variables go together. By structuring information into objects, such relations can be discovered automatically.

Methods. A *MethodNode* represents a procedure associated with an object. The method signature are two arrays of input and output arguments. Every argument is described, similar to variables, by a data type, value rank and array dimensions. Several objects may reference the same *MethodeNode*. The request message of the *Call* service contains the *NodeId* of the method and the *NodeId* of the object as context along with the input arguments.

Object Types. Additional structure is imposed on objects via *ObjectTypes* that define mandatory and optional member objects, methods and variables. Object types are the smallest common denominator of the entities they represent. Then, programs are written against the object type definition to reduce the amount of integration work required for every instance. Figure 2 shows an example object type hierarchy. The object type *Pump* is derived from the generic *FieldDevice*. It specifies two variables and two methods for interacting with the physical pump it represents. Every object needs one *hasTypeDefinition* reference to an object type. The default object type, *BaseObjectType*, is at the root of the object type hierarchy and defines no members. Similar to the previously discussed node classes for describing a type, object types may be abstract, in which case they cannot be instantiated. Advanced concepts of object-orientation, such as multiple inheritance are not ruled out in the standard, but their use is currently underspecified.

Views. Information models can be large. Views give access to a subset of nodes to reduce complexity. That is, the information model can be browsed in reference to a *ViewNode*. For example, in an information model describing a chemical plant, a view may filter out all nodes not pertaining to an electrical device.

3   Support for Big-Data Volume, Velocity and Variety

Big-Data stands for the use of huge datasets for analytics. The term is often defined in reference to the data volume, velocity and variety. In this section, we detail the features of OPC UA that render it useful for data storage and transfer according to the three "V".

3.1   Volume: Working with Large Data Sets

In order to deal with large information models, OPC UA defines several mechanisms for focused access to a subset of the data and its piecewise transfer.

Historical Data Access. Many "big" data sets deal with time-series data. In OPC UA, historical values of node attributes can be retrieved with the *HistoryRead* service and changed with the *HistoryUpdate* service. Most server implementations have this feature enabled only for the value attribute of select *VariableNodes*. The possibilities range from specialized databases for time-indexed values to a simple ring-buffer for storing values directly in RAM.

Reading with an Index Range. The value attribute of a *VariableNode* may be a (multi-dimensional) array. These can naturally become quite large so that transferring the entire array should be avoided if possible. The *Read* and *Write* service for accessing node attributes optionally contain an index-range in string encoding, such as "1:3, 5, 0:5", to access specific entries in a (multi-dimensional) array.

Chunking. Since the response to a *Read* request may be larger than the Maximum Transmission Unit (MTU) of the underlying network, OPC UA uses chunking where messages are split into several packets and reassembled on the receiving end. It is not uncommon to see maximum message sizes that are several hundred megabyte long. This is of course limited on embedded systems with less available working memory.

Resumable Services. Some services, such as *Browse* may have an unexpectedly large response. In the case of *Browse*, clients can limit the number of returned node references. The operation can be resumed with a *BrowseNext* call so that references are not missed.

File Transfer. Since the advent of Unix, huge data blobs are often stored in files in a folder hierarchy. A special *ObjectType* called *FileType* is used to expose such a filesystem folder hierarchy and common operations of the underlying filesystem (open, read, skip to position, write, copy file, ...) in OPC UA.

3.2 Velocity: Handling High Throughput

OPC UA was designed to work in conditions where only few resources are available for computation and data transfer. Together with asynchronous messaging enabling an event-oriented programming style, OPC UA can be scaled to handle a large throughout of data.

Binary Protocol. OPC UA defines a binary encoding for the protocol type system in which request and response messages are defined. This leads to a small size of messages. Encoding messages in XML and transport over HTTP/SOAP is alternatively possible. But the increase in message size is reported to be over an order of magnitude.

Batch Operations. Most services defined in OPC UA are batched. For example, a *ReadRequest* contains an array of the nodes and their attributes to be read. (See also Listing 1 for a request message containing an array of actual read operations.) This generally leads to fewer larger packets that are sent instead of many smaller ones. On packet-switched networks, this leads to a better use of the available capacity for data transport.

Asynchronous messaging. The asynchronous design of OPC UA, where message responses may delayed or occur in a different ordering, encourages an event-oriented programming style. The advantages of this can be seen, for example in the performance increase of modern webservers, such as nginx, versus their counterparts from the 90s, such as Apache. See also Sect. 4 for the impact on the design of the *open6251* implementation.

Subscriptions. Polling, i.e. repeated reading of values, is a highly inefficient way to track runtime value changes and events. Push-notification can drastically reduce the number of exchanged messages when only notifications occur not very often, but need to be communicated in a timely manner. Even though OPC UA strictly adheres to the request/response pattern, push-notification is one of its core concepts. Since a server can send responses asynchronously and out-of-order, requests for the *Publish* service are queued up for sending notifications at a later time. OPC UA differentiates between so-called *MonitoredItems* and *Subscriptions*. *MonitoredItems* either track a data value change with a fixed sampling interval (intermediate values are lost if the sampling interval is too long) or listen for *Events* (every *Event* generates a notification). Notifications are fed into the publication queue of a *Subscription*. Every *Subscription* has a publication interval. Only when the publication interval times out, queued notifications are sent out in a single message to reduce the network load. If no notifications have been accumulated, a heartbeat message is sent after some number of publication interval without activity. Extensive measures have been taken that clients can retrieve a full history of the notifications created by the server. The receipt of a *Publish* response needs to be acknowledged by the client in one of the following requests. Until then, old notifications are kept in the server's working memory. If, for example, communication is interrupted and the *SecureChannel* breaks down, the client can reestablish the old *Session* on a new *SecureChannel* and retrieve non-acknowledged notifications via the *Republish* service.

3.3   Variety: Handling Heterogeneous Data

OPC UA as a NoSQL Database. Relational databases rely on a fixed schema for each table. They are however not suited for storing unstructured data (not considering ad-hoc encoding as string values). In recent years, so-called NoSQL databases have seen increased use. An OPC UA server can be seen as a special type of NoSQL database combining features from object- and graph-databases. The difference to databases based on Codd's relational algebra [3] is that triple-relations are used [2] as the underlying data structure. Instead of SQL, the *Browse* service is used to discover the information at runtime. The *Query* service is used once the data's meta-model is sufficiently understood and clear search criteria can be defined. But even though OPC UA information models are based on triple-relations, the OPC UA query mechanism does not offer the inferential power of SPARQL endpoints used for RDF-based [12] semantic models.

Flexible Protocol Type System. On todays Internet, ad-hoc data encoding with JSON is quite popular since the format is very flexible. Some of the NoSQL databases, such as MongoDB, leverage this flexibility and use JSON as the core format of schema-less data base entries. Even though OPC UA defines a protocol type system, servers and clients do not have to know about the transferred data literals a-priori. First, data types can be introspected based on their *DataTypeNode* in the server's information model. Second, the binary protocol defines special way to transfer complex data types. The so-called *ExtensionObject* data type is used to encode values that are not one of the 25 built-in types. In the binary encoding, the *ExtensionObject* begins with the *NodeId* of the content's data type and the length of the following binary encoding. This ensures that the receiving end can decode the message even if it has no notion about a portion of its content. Servers can then store the value as a binary blob and let clients interact with it.

4   Design of the open62541 Implementation to Support Semantic Interoperability at Scale

open62541 (http://open62541.org) is an open source implementation of OPC UA. It is a library written in the common subset of the C and C++ language and provides functionality to implement dedicated OPC UA clients and servers, or to integrate OPC UA-based communication into existing applications. *open62541* was started in the beginning of 2014 and driven by several universities and research organizations with the joint requirement of being in full control of the communication stack for research projects. Since then, large parts of the specification have been implemented and pass the official Conformance Testing Tools (CTT) of the OPC Foundation. *open62541* currently implements the core OPC UA communication stack as well as the server and client SDK in about 15,000 lines of code (not counting generated code). In the remainder of this section, we discuss the design choices made for *open62541* for scaling from embedded applications to large multi-core servers for big-data operations.

Protocol Type System. At the core of the stack lies the protocol type system. We differentiate between the protocol type system and typing mechanisms used for information models. Note however, that every data type defined in the protocol type system is identified by *NodeId* of a corresponding *DataTypeNode*. Only the values defined in terms of the protocol type system can be encoded in binary messages to be sent over the network. The OPC UA protocol defines 25 built-in data types and three ways of combining them into higher-order types: arrays, structures and unions. In *open62541*, the built-in data types are defined manually. All other data types are generated from standard XML definitions. Their exact definitions can be looked up at https://opcfoundation.org/UA/schemas/ Opc.Ua.Types.bsd.xml. Listing 1 shows an example data type in its XML definition and the resulting structure in the C programming language. In order to reduce the binary size, *open62541* does not define extra handling functions for every data type. Instead, a struct with the type description is handed over to generic functions, e.g., for binary encoding. The encoding function is optimized for minimal resource consumption. Beside speed optimization, it is possible to encode messages that are bigger than the network buffer. For this, a callback is triggered when the end of the encoding buffer is reached. Then, the current buffer is sent out and reset, so that the encoding function can continue where it left off. This saves one sweep over every data type that would otherwise be required to determine the binary encoding length prior to allocating a buffer. The encoding has generic fallbacks that work on any processor architecture, including the various possibilities for endianness and non IEEE-754 floating point number representations.

```xml
<opc:StructuredType Name="ReadRequest" BaseType="ua:ExtensionObject">
  <opc:Field Name="RequestHeader" TypeName="tns:RequestHeader" />
  <opc:Field Name="MaxAge" TypeName="opc:Double" />
  <opc:Field Name="TimestampsToReturn" TypeName="tns:TimestampsToReturn" />
  <opc:Field Name="NoOfNodesToRead" TypeName="opc:Int32" />
  <opc:Field Name="NodesToRead" TypeName="tns:ReadValueId"
          LengthField="NoOfNodesToRead" />
</opc:StructuredType>
```

```c
typedef struct {
    UA_RequestHeader requestHeader;
    UA_Double maxAge;
    UA_TimestampsToReturn timestampsToReturn;
    size_t nodesToReadSize;
    UA_ReadValueId *nodesToRead;
} UA_ReadRequest;
```

Listing 1. Structured type for a service request in the XML definition and the generated C-structure in open62541.

Network Connection Handling. Network connections are made up of three layers. Lowest is the raw TCP connection. For this, a plugin API has been defined so that the networking layer can be exchanged for non-POSIX targets. On top of every TCP connection, a *SecureChannel* is established. (The security policy "None" defines a *SecureChannel* without encryption or signing.) A *Session* requires users to authenticate. Their credentials, such as username and password, are verified by the user that can also assign custom data to the session. Later on, this data is passed to the access rights management function when the user interacts with the information model. *Sessions* are stateful and may outlive the current *SecureChannel*. For this, they are rebound to a new *SecureChannel*. This is important for users who do not want to loose subscription notifications.

Event-Oriented Architecture. OPC UA strictly adheres to the request/response pattern where only clients can send requests. However, responses are asynchronous. That is, servers may respond to requests in a different ordering and may delay responses. This encourages a non-blocking style of programming, greatly enhancing the responsiveness of applications. This led us to adopting an event-based architecture for *open62541*. Messages are retrieved from the network layer and added to a dispatch queue. There, independent worker threads dequeue events for processing (see Fig. 3). In the single-core case, dispatching an event is synonymous with processing it right away. Similar architectures have led to huge increases in latency and throughput for webservers [15]. Also, the event-oriented architecture enables the efficient handling of repeated callbacks at different cycle intervals. With *open62541*, it is possible to establish tens of thousands of *MonitoredItems* with a few millisecond sampling interval each. Their execution is distributed across processor cores. In the next paragraph, we detail why this does lead to contention points that limit scaling.

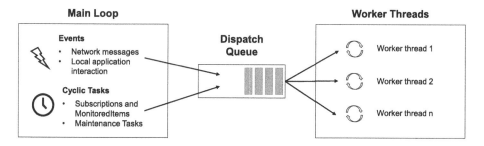

Fig. 3. Event-based architecture

Immutable Data Structures and Lockless Parallelism. Since the processor clock-speeds reached their current limit around 2005, multicore operations have become the leading paradigm for scaling applications. However, the achievable speed-up is limited by Amdahl's Law [9] according to the portion of the rate-limiting code that does not support parallelism and needs to be protected by locks. To overcome these limits, recent processors support atomic operations, such as CAS (compare-and-switch) [13]. Atomic operations make consistency guarantees across all processor cores and memory caching levels. In the case of CAS, a pointer is replaced only when the current value still points to the same memory address. The NodeStore in *open62541* is implemented as a hashmap where OPC UA nodes are indexed with their *NodeId*. In multicore operations, nodes cannot be edited and must be replaced in their entirety. This can be achieved with atomic operations. The question remains when old versions can be safely removed without expensive synchronisation. Currently, *open62541* uses the RCU (Read, Copy, Update) mechanism originally developed to scale concurrent operations in the Linux kernel [4]. To free other shared data structures, for example a *Session*, delayed callbacks are used. When the *Session* is closed, it is invalidated and all pointers to the session are removed. But freeing the memory is delayed until all previously dispatched events, where the *Session* may still be references, have completed to protect from data races [7].

Service Sets. The 37 standard OPC UA services are organized into service sets. Some services, for example to establish a *Session* can be accessed only over the network. Most services however are exposed via the user-visible API to the server developer. In fact, services are the only way for users to interact with the server's information model. That reduces the complexity as there is a single point where, e.g., node attributes can be set. Furthermore, any functionality that is implemented for the local user, such as adding a new object instance, is automatically available also over the network.

Consider now the following microbenchmark of the *Read* service as an indication of the effectiveness of the services (making heavy use of the NodeStore in the background) and the encoding function. Decoding the read-request, processing the service and encoding the response can be repeated more than 1,000,000 times per second on a laptop computer (running an Intel i7-3520M at 2.9 GHz).

Thus, less than 3000 processor cycles are required for this very common service.[1] This however excludes the effect of networking itself and the necessary context switches into the operating system kernel.

5   Conclusion

In this paper, we discussed the use of OPC UA, a network protocol commonly used in the Industrial Internet of Things, to support semantic interoperability and connectivity in big-data scenarios. We then discussed the architecture of the *open62541* OPC UA implementation and the measures taken to ensure scalability. In recent years, several open source implementations of OPC UA have been developed for the most common software development platforms. That is OPC UA is becoming a readily available technology. With its features for semantic interoperability and large-scale application scenarios, OPC UA brings the learnings for the integration of large-scale cyber-physical systems from the industrial automation community to a wider audience. Our hope is that a fruitful exchange of ideas will ensue.

References

1. Arai, T., Aiyama, Y., Maeda, Y., Sugi, M., Ota, J.: Agile assembly system by 'plug and produce'. CIRP Ann. Manuf. Technol. **49**(1), 1–4 (2000)
2. Chen, P.P.S.: The entity-relationship model–toward a unified view of data. ACM Trans. Database Syst. **1**(1), 9–36 (1976)
3. Codd, E.F.: A relational model of data for large shared data banks. Commun. ACM **13**(6), 377–387 (1970)
4. Desnoyers, M., McKenney, P.E., Stern, A.S., Dagenais, M.R., Walpole, J.: User-level implementations of read-copy update. IEEE Trans. Parallel Distrib. Syst. **23**(2), 375–382 (2012)
5. Frey, C.W.: Diagnosis and monitoring of complex industrial processes based on self-organizing maps and watershed transformations. In: 2008 IEEE International Conference on Computational Intelligence for Measurement Systems and Applications, pp. 87–92. IEEE (2008)
6. Gaj, P., Jasperneite, J., Felser, M.: Computer communication within industrial distributed environment - a survey. IEEE Trans. Ind. Inf. **9**(1), 182–189 (2013)
7. Hart, T.E., McKenney, P.E., Brown, A.D., Walpole, J.: Performance of memory reclamation for lockless synchronization. J. Parallel Distrib. Comput. **67**(12), 1270–1285 (2007)
8. Heiler, S.: Semantic interoperability. ACM Comput. Surv. (CSUR) **27**(2), 271–273 (1995)
9. Hill, M.D., Marty, M.R.: Amdahl's law in the multicore era. Computer **41**(7), 33–38 (2008)
10. Hitzler, P., Krotzsch, M., Rudolph, S.: Foundations of Semantic Web Technologies. CRC Press, Boca Raton (2009)

---

[1] The measurement code is accessible under https://github.com/open62541/open62541/blob/0.2/examples/server_readspeed.c.

11. IEC 62541. OPC Unified Architecture Part 1–10, Release 1.0 (2010)
12. McBride, B.: The resource description framework (RDF) and its vocabulary description language RDFS. In: Staab, S., Studer, R. (eds.) Handbook on Ontologies. International Handbooks on Information Systems, pp. 51–65. Springer, Heidelberg (2004)
13. McKenney, P.E.: Is parallel programming hard, and if so, what can you do about it? Technical report (2011). https://www.kernel.org/pub/linux/kernel/people/paulmck/perfbook/perfbook.html
14. Niggemann, O., Biswas, G., Kinnebrew, J.S., Khorasgani, H., Volgmann, S., Bunte, A.: Data-driven monitoring of cyber-physical systems leveraging on big data and the internet-of-things for diagnosis and control. In: Proceedings of the 26th International Workshop on Principles of Diagnosis. ACM (2015)
15. Pariag, D., Brecht, T., Harji, A., Buhr, P., Shukla, A., Cheriton, D.R.: Comparing the performance of web server architectures. In: ACM SIGOPS Operating Systems Review, vol. 41, pp. 231–243. ACM (2007)
16. Pfrommer, J., Stogl, D., Aleksandrov, K., Escaida Navarro, S., Hein, B., Beyerer, J.: Plug & produce by modelling skills and service-oriented orchestration of reconfigurable manufacturing systems. at-Automatisierungstechnik **63**(10), 790–800 (2015)

Author Index

Printed in the United States
By Bookmasters